CANCER ETIOLOGY, DIAGNOSIS AND TREATMENTS SERIES

NEW QUINOLONES WITH POTENTIAL ANTI-MRSA ACTIVITY

CANCER ETIOLOGY, DIAGNOSIS AND TREATMENTS SERIES

Cell Apoptosis and Cancer
Albina W. Taylor (Editor)
2007. ISBN: 1-60021-506-8

Chronic Lymphocytic Leukemia Research Focus
Chadi Nabhan (Editor)
2007. ISBN: 1-60021-526-2

Cervical Cancer Research Trends
Eleanor P. Bankes (Editor)
2007. ISBN: 1-60021-648-x

Lung Cancer in Women
Varetta N. Torres (Editor)
2008. ISBN: 1-60021-659-5

Lung Cancer in Women
Varetta N. Torres (Editor)
2008. ISBN: 978-1-60692-765-6 (Online book)

Cancer Research at the Leading Edge
Ignatius K. Martakis (Editor)
2008. ISBN: 1-60021-728-1

Chronic Lymphocytic Leukemia: New Research
Inès B. Moreau (Editor)
2008. ISBN: 978-1-60456-081-7

Cancer and Stem Cells
Thomas Dittmar and Kurt S. Zander (Editors)
2008. ISBN: 978-1-60456-478-5

Cancer and Stem Cells
Thomas Dittmar and Kurt S. Zander (Editors)
2008. ISBN: 978-1-61668-044-7 (Online Book)

Cancer Prevention Research Trends
Louis Braun and Maximilian Lange (Editors)
2008. ISBN: 978-1-60456-639-0

Clinical, Genetic and Molecular Precursor Features in Colorectal Neoplasia
Kjetil Søreide and Håvard Søiland (Editors)
2008. ISBN: 978-1-60456-714-4

Drug Resisant Neoplasms
Ethan G. Verrite (Editor)
2009. ISBN: 978-1-60741-255-7

Handbook of Prostate Cancer Cell Research: Growth, Signalling and Survival
Alan T. Meridith (Editor)
2009. ISBN: 978-1-60741-954-9

Human Polyomaviruses: Molecular Mechanisms for Transformation and their Association with Cancers
Ugo Moens, Marijke Van Gheule and Mona Johannessen
2009. ISBN: 978-1-60692-812-7

Anticancer Drugs: Design, Delivery and Pharmacology
Peter Spencer and Walter Holt (Editors)
2009. ISBN: 978-1-60741-004-1

Anticancer Drugs: Design, Delivery and Pharmacology
Peter Spencer and Walter Holt (Editors)
2009. ISBN: 978-1-60876-629-1 (Online Book)

Cancer Biology: An Updated Global Overview
Tarek H. EL-Metwally
2009. ISBN: 978-1-60876-193-7

Molecular Therapy of Breast Cancer: Classicism Meets Modernity
Marc Lacroix
2009. ISBN: 978-1-60741-593-0

Molecular Therapy of Breast Cancer: Classicism Meets Modernity
Marc Lacroix
2009. ISBN: 978-1-60876-726-7 (Online Book)

Aromatase Inhibitors: Types, Mode of Action and Indications
Jean R. Lamonte (Editor)
2009. ISBN: 978-1-60741-711-8

Aromatase Inhibitors: Types, Mode of Action and Indications
Jean R. Lamonte (Editor)
2009. ISBN: 978-1-61668-665-9

Small Cell Carcinomas: Causes, Diagnosis and Treatment
Jonathon G. Maldonado and Mikayla K. Cervantes (Editors)
2009. ISBN: 978-1-60741-787-3

Multiple Myeloma: Symptoms, Diagnosis and Treatment
Milen Georgiev and Evgeni Bachev
2009. ISBN: 978-1-60876-108-1

Viral Cancers: Cytologic Tools in Diagnosis and Management
Dilip K. Das
2010. ISBN: 978-1-60876-402-0

Nose and Viral Cancer: Etiology, Pathogenesis and Treatment
Aloisio Medeiros and Carlitos Veloso (Editors)
2010. ISBN: 978-1-60741-735-4

Karyogamic Theory of Cancer Cell Formation from the View of the XXI Century
G.Gogichadze and T.Gogichadze
2010. ISBN: 978-1-60876-386-3

Aggressive Breast Cancer
Regina H. DeFrina (Editor)
2010. ISBN: 978-1-60876-881-3

Human Papillomavirus (HPV) Involvement in Esophageal Carcinogenesis
Kari Syrjänen
2010: ISBN: 978-1-60876-211-8

New Quinolones with Potential Anti-MRSA Activity
Saeed Emami
2010. ISBN: 978-1-60876-736 6

PSA and Prostate Cancer
Jake A. Saylor and Lionel B. Michaels (Editors)
2010. ISBN: 978-1-60876-895-0

Pentacyclic Triterpenes as Promising Agents in Cancer
Jorge A. R. Salvador (Editor)
2010. ISBN: 978-1-60876-973—5

Breast Cancer: Causes, Diagnosis and Treatment
Martin E. Romero and Louis M. Dashek (Editors)
2010. ISBN: 978-1-60876-463-1

Gastric Cancer: Diagnosis, Early Prevention, and Treatment
Victor D. Pasechnikov (Editor)
2010. ISBN: 978-1-61668-313-9

New Approaches in the Treatment of Cancer
Dra. Ma. Del Camen Mejia Vazquez, Samuel Navarro
2010. ISBN: 978-111-61668-361-0

Handbook of Skin Care in Cancer Patients
Pierre Vereecken, Ahmad Awada, Jules Bordet (Editors)
2010. ISBN: 978-1-61668-419-8

MicroRNAs in Breast Cancer
Marc Lacroix
2010. ISBN: 978-1-61668-438-9

MicroRNAs in Breast Cancer
Marc Lacroix
2010. ISBN: 978-1-61668-498-3 (Online Book)

CANCER ETIOLOGY, DIAGNOSIS AND TREATMENTS SERIES

NEW QUINOLONES WITH POTENTIAL ANTI-MRSA ACTIVITY

SAEED EMAMI

Nova Science Publishers, Inc.
New York

Copyright © 2010 by Nova Science Publishers, Inc.

All rights reserved. No part of this book may be reproduced, stored in a retrieval system or transmitted in any form or by any means: electronic, electrostatic, magnetic, tape, mechanical photocopying, recording or otherwise without the written permission of the Publisher.

For permission to use material from this book please contact us:
Telephone 631-231-7269; Fax 631-231-8175
Web Site: http://www.novapublishers.com

NOTICE TO THE READER

The Publisher has taken reasonable care in the preparation of this book, but makes no expressed or implied warranty of any kind and assumes no responsibility for any errors or omissions. No liability is assumed for incidental or consequential damages in connection with or arising out of information contained in this book. The Publisher shall not be liable for any special, consequential, or exemplary damages resulting, in whole or in part, from the readers' use of, or reliance upon, this material.

Independent verification should be sought for any data, advice or recommendations contained in this book. In addition, no responsibility is assumed by the publisher for any injury and/or damage to persons or property arising from any methods, products, instructions, ideas or otherwise contained in this publication.

This publication is designed to provide accurate and authoritative information with regard to the subject matter covered herein. It is sold with the clear understanding that the Publisher is not engaged in rendering legal or any other professional services. If legal or any other expert assistance is required, the services of a competent person should be sought. FROM A DECLARATION OF PARTICIPANTS JOINTLY ADOPTED BY A COMMITTEE OF THE AMERICAN BAR ASSOCIATION AND A COMMITTEE OF PUBLISHERS.

LIBRARY OF CONGRESS CATALOGING-IN-PUBLICATION DATA

New quinolones with potential anti-MRSA activity / Saeed Emami.
    p. ; cm.
Includes bibliographical references and index.
ISBN 978-1-60876-736-6 (softcover)
1. Quinolone antibacterial agents. 2. Staphylococcus aureus infections--Chemotherapy. I. Title.
[DNLM: 1. Quinolones--pharmacology. 2. Anti-Bacterial Agents--pharmacology. 3. Drug Resistance, Multiple, Bacterial. 4. Methicillin-Resistant Staphylococcus aureus--drug effects. 5. Staphylococcal Infections--drug therapy. QV 250 E56n 2009]
RM666.Q55.E43 2009
616.9'2061--dc22
                          2009045937

Published by Nova Science Publishers, Inc. † New York

CONTENTS

Preface		xi
Introduction		xiii
Chapter 1	History of Quinolones Development	1
Chapter 2	Mechanism of Action	7
Chapter 3	Mechanisms of Quinolone Resistance in *S. aureus*	11
Chapter 4	General Structural Features (Structure-activity Relationships)	17
Chapter 5	Newer Quinolones with Potential Anti-MRSA Activity	27
Summary		85
References		87
Index		105

PREFACE

Quinolone antibacterial agents are among the most attractive drugs in the field of anti-infective chemotherapy. Quinolones target the essential bacterial type II topoisomerases, which are involved in DNA replication and metabolism. Despite a large number of quinolones approved for the treatment of bacterial infections, there have been progressive efforts for the discovery of new quinolones to overcome the problem of growing bacterial resistance. Over the years, there has been growing concern surrounding the increased prevalence of drug resistance in *Staphylococcus aureus*, especially methicillin-resistant *Staphylococcus aureus* (MRSA). MRSA used to be susceptible to the fluoroquinolones when they were introduced in the early 1980s but a rapid emergence of resistance to these drugs was observed soon after the increasing use of fluoroquinolones from the mid 1980s. Progressive efforts have been made to optimize the biological activity of the quinolones against *S. aureus* and drug-resistant strains including MRSA. Several newly developed quinolones and related compounds exhibit enhanced activity against *S. aureus* and have proved useful against MRSA strains and are under rapid preclinical and clinical development.

This book will discuss the structural requirements of quinolones for anti-MRSA activity and introduce the newly developed quinolones to manage MRSA infections, their in-vitro effectiveness based on recent studies and structure-activity relationships. The first three chapters cover concisely the history of quinolones development, their mechanism of action and mechanisms of quinolone resistance in *S. aureus*. The fourth chapter describes general structural features of quinolones for anti-staphylococcal activity and their structure-activity relationships. The final chapter (chapter 5) is devoted to review several newly developed quinolones and related compounds including 6-fluoroquinolones, 6-fluoronaphthyridones, 6-desfluoroquinolones, quinazolinediones and 4-quinolizinones (2-pyridones) with enhanced activity against *S. aureus* which have proved useful against MRSA. The structures and activities of these categories of quinolones will be discussed in this chapter.

I earnestly hope that this book will provide the kind of information that will interest those who work or plan to begin work in this captivating area of quinolone antibacterial agents.

Iran, Sari, June 2009   Saeed Emami

INTRODUCTION

Methicillin-resistant *Staphylococcus aureus* (MRSA) is a multidrug-resistant pathogen that has historically associated with hospitals and healthcare facilities. Over the years, there has been growing concern surrounding the increased prevalence of drug resistance in *S. aureus*, especially hospital-associated MRSA (HA-MRSA), which limits drug options [1]. Until late in 1997, vancomycin was the only agent available with reliable activity against serious MRSA infections. Other agents, including rifampin, novobiocin, daptomycin, teicoplanin, tigecycline, streptogramins, oxazolidinones and quinolones have been used to treat patients infected with MRSA in an ongoing effort to expand treatment options [2].

The quinolones are synthetic antibacterial agents, first used clinically in the 1960s with the introduction of nalidixic acid. The utility of nalidixic acid was restricted to Gram-negative bacteria. In the 1960s, MRSA had only recently been described; quinolone compounds with activity against staphylococci were not available until the introduction of fluoroquinolones, such as ciprofloxacin and ofloxacin, two decades later. MRSA used to be susceptible to the fluoroquinolones when they were introduced in the early 1980s but a rapid emergence of resistance to these drugs was observed soon after the increasing use of fluoroquinolones from the mid 1980s. Improvement to the chemical structure of early quinolones, based on the addition of a fluorine atom at the C-6 position and modification of the C-7 position, dramatically enhanced antibacterial spectrum and improved pharmacokinetics. Much of the improved potency of modern quinolones against staphylococci has been achieved by tinkering with the N-1, C-7, and C-8 substituents on the quinolone ring system.

The widespread emergence of antibiotic resistance Gram-positive bacteria has stimulated the search for effective new quinolone antibacterial agents in recent years. Generally, the newer quinolones retain the broad spectrum activity of the currently used quinolones, such as levofloxacin and moxifloxacin, and also display improved activity against *S. aureus*, including some MRSA isolates. Several new agents have proved useful against MRSA strains and are under rapid preclinical and clinical development.

The aim of this book is to focus on the pharmacophoric requirements of quinolones for anti-staphylococcal activity and the newly developed quinolones to manage MRSA infections, their in-vitro effectiveness based on recent studies and structure-activity relationships.

Chapter 1

HISTORY OF QUINOLONES DEVELOPMENT

Quinolones are synthetic drugs developed through structural modification of the 'quinoline' skeleton. In 1962, during the process of synthesis of chloroquine (an antimalarial agent), a napththyridone (8-azaquinolone) derivative, nalidixic acid (Figure 1), was discovered which possessed bactericidal activity [3]. The use of nalidixic acid was limited because of its narrow spectrum, low serum levels, and toxicity issues, but it regained attention in the 1980s for the treatment of diarrhea and urinary tract infections following the development of resistance in *Shigella* and *Escherichia coli* to other classes of antibiotics used at that time [4]. Thereafter, novel compounds of this family, such as oxolinic acid and pipemidic acid (Figure 1), were synthesized and introduced into clinical practice, although the clinical indication for these quinolones still remained only for urinary tract infections.

The introduction of a piperazin-1-yl moiety at 7-position (to provide pipemidic acid) did improve the activity against Gram-negatives, broadening the spectrum to include *Pseudomonas* spp. Pipemidic acid also had some activity against Gram-positive species. This piperazine ring has subsequently been shown to increase the ability of the quinolones to penetrate the bacterial cell wall, thus enhancing activity. The first compound with a fluorine atom at position 6 was flumequine (patented in 1973) and this compound gave the first indications that activity against Gram-positive organisms (e.g. *S. aureus*) could be improved in this class [5]. Following the introduction of flumequine, the second generation of quinolones had the major feature of a fluorine substituent at position 6, which increased activity markedly. These early compounds were most potent against Gram-negative organisms; thus their activity against *Streptococcus pneumoniae* was too marginal to warrant clear indications for use in the treatment of respiratory tract infections, and the emergence of resistance soon reduced their potential against *S. aureus* [6, 7].

The real breakthrough came with the combination of these two features in norfloxacin (Figure 2), a 6-fluoroquinolone with a piperazine ring at position 7. Although norfloxacin does have activity against a broad spectrum of Gram-negatives and has some Gram-positive activity, the combination of its pharmacokinetic profile and activity were still not adequate for systemic use [8-10].

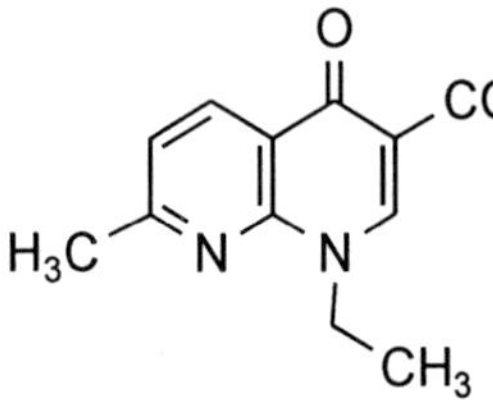

Nalidixic acid

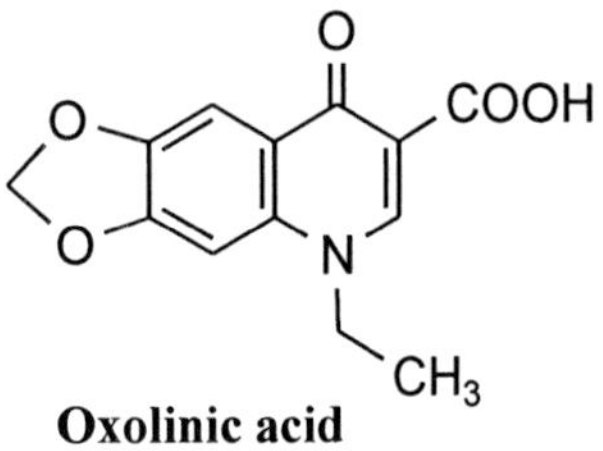

Oxolinic acid

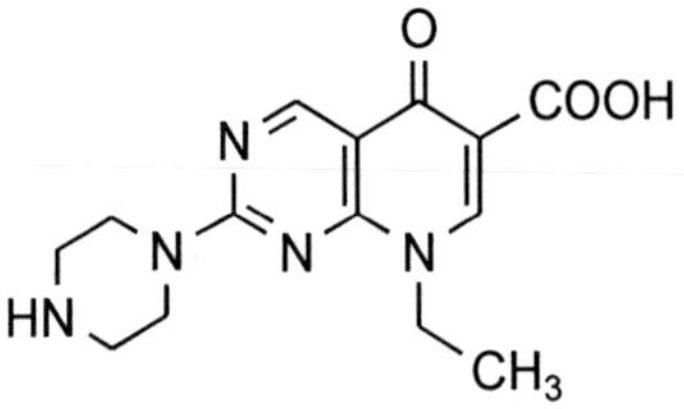

Pipemidic acid

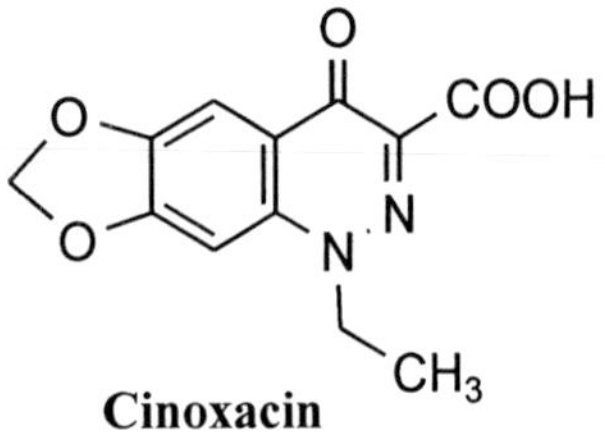

Cinoxacin

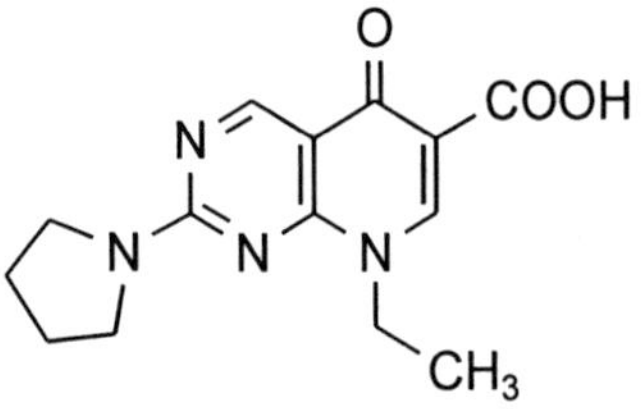

Piromidic acid

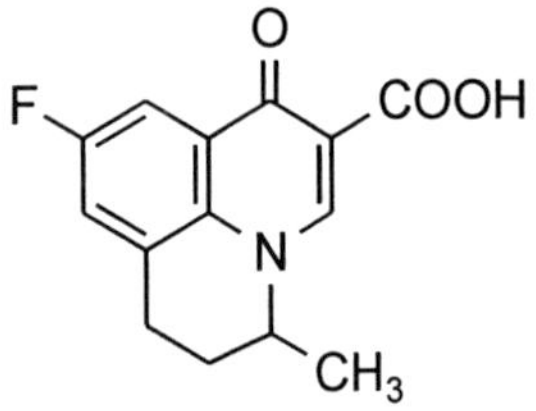

Flumequine

Figure 1. Older quinolones.

Generation IIa:

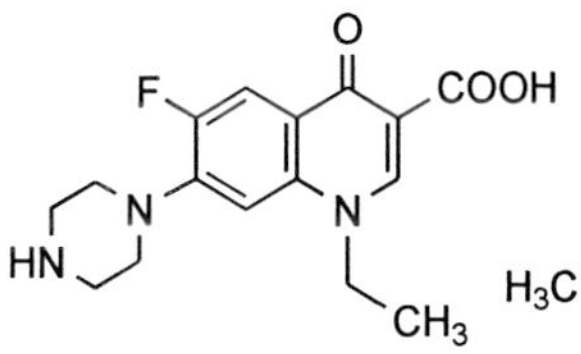

Norfloxacin

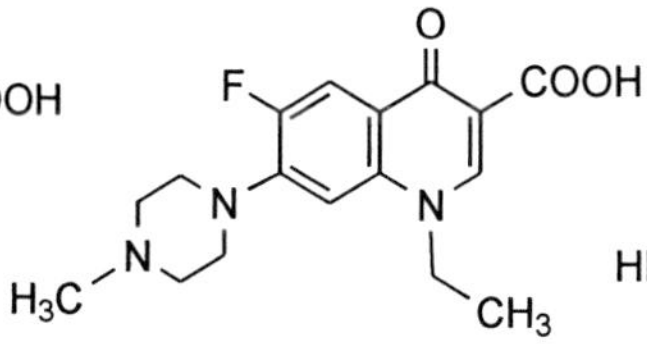

Pefloxacin

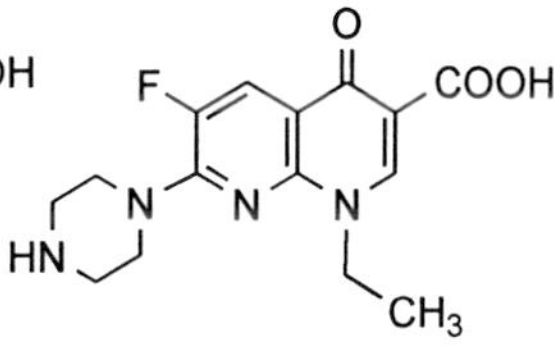

Enoxacin

Fleroxacin

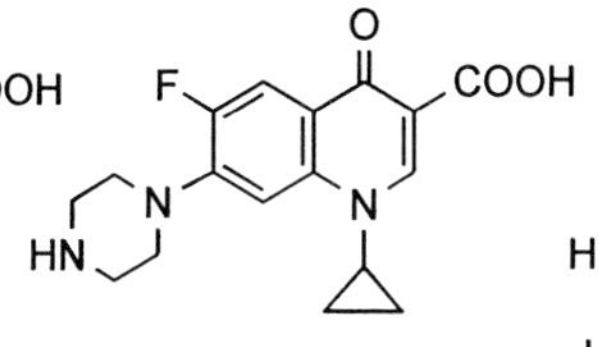

Ciprofloxacin

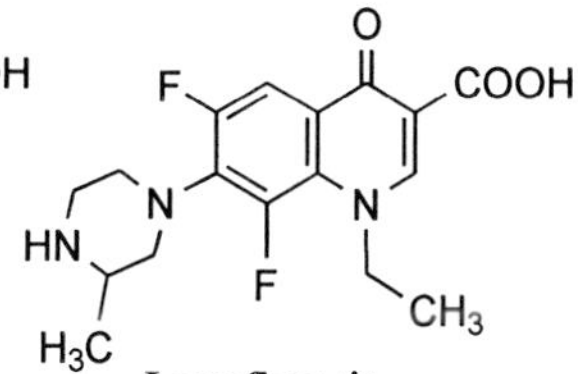

Lomefloxacin

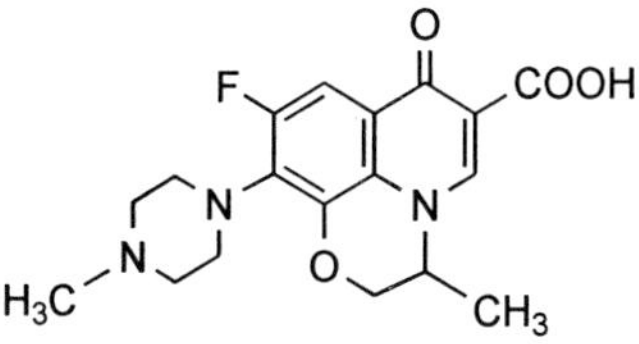

Ofloxacin

Generation IIb:

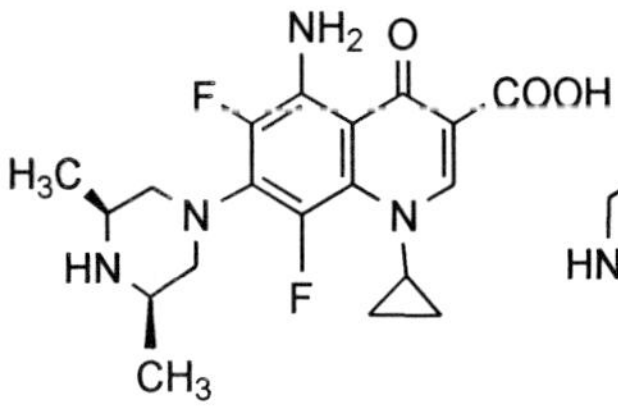

Sparfloxacin

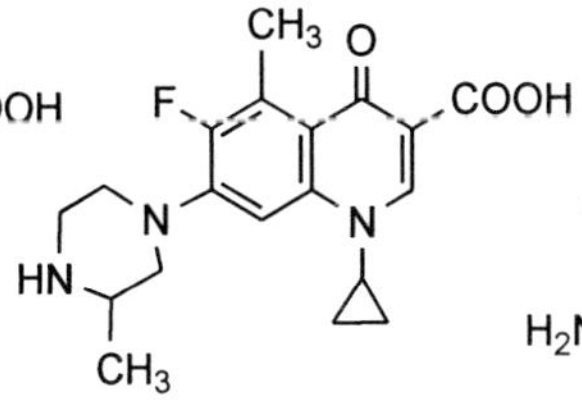

Grepafloxacin

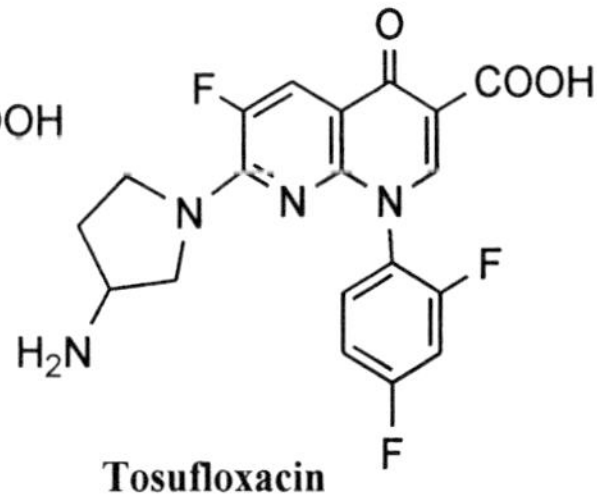

Tosufloxacin

Figure 2. Second-generation of quinolones (6-fluoroquinolones).

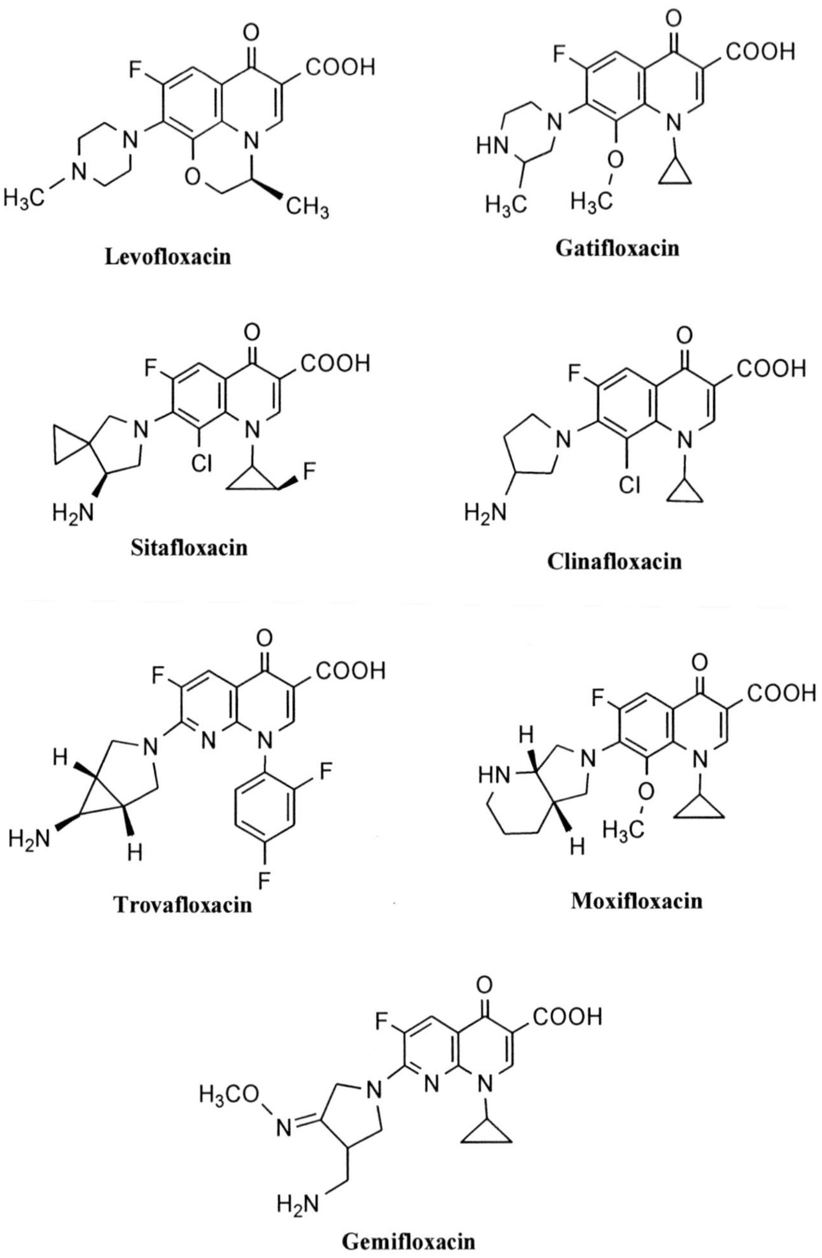

Figure 3. Third-generation quinolones with improved anti-staphylococcal activity.

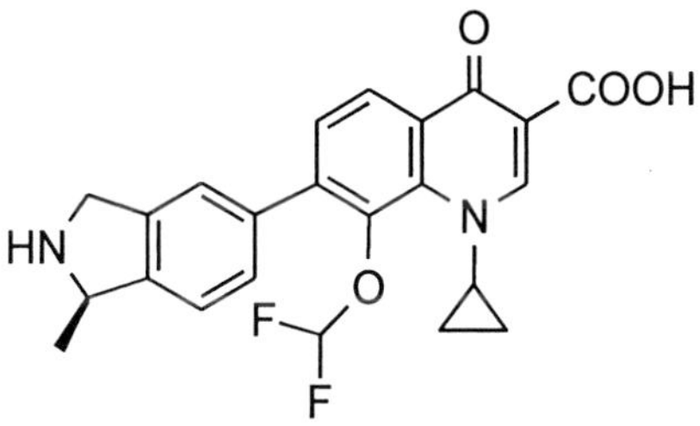

Garenoxacin

Figure 4. Structure of garenoxacin; a representative 6-desfluoroquinolone.

Following the introduction of norfloxacin and several related compounds including pefloxacin, enoxacin, fleroxacin and lomefloxacin (Figure 2), the substitution of a cyclopropyl group at position N-1, as in ciprofloxacin, gave a marked improvement in activity. Ofloxacin exemplify another structural modification in the form of a bridging ring between the N-1 and position 8, which also gives improved activity [5]. The advantages of these drugs are that their spectrum includes Gram-positive organisms as well as Gram-negatives, and that they are well absorbed from the gastrointestinal tract, providing adequate blood levels to allow their use for systemic infections. However, their activity against Gram-positive cocci is inadequate for the treatment of respiratory tract infections. Similarly, although the development of resistance to these agents is clearly less of a problem than that with nalidixic acid, in some species, particularly *S. aureus* and *Pseudomonas aeruginosa*, resistance to this group has become a problem [11].

The quinolones were further optimized by the addition of different functional groups to the N-1, C-5, C-7 and C-8 positions of their respective basic molecules. A group of compounds showing more dramatic improvements are sparfloxacin, grepafloxacin, tosufloxacin (Figure 2), levofloxacin, gatifloxacin, clinafloxacin, sitafloxacin, moxifloxacin, trovafloxacin and gemifloxacin (Figure 3). They all show good activity against most Gram-positives, with clinafloxacin, gemifloxacin and sitafloxacin having the best activity [12, 13].

Optimising all other substituents has permitted the removal of the fluorine atom at position 6 (which has been claimed to be involved in genotoxicity and central nervous system defects possibly involved in genotoxicity), giving rise to the new generation of quinolones; termed des-fluoroquinolones, with garenoxacin as its first representative (Figure 4). The extensive research efforts have enabled a better definition of the structural moieties or elements around the basic pharmacophore that offer the best combination of clinical efficacy, reduced resistance selection, and safety.

Chapter 2

MECHANISM OF ACTION

Quinolones exert their antibacterial action via inhibition of homologous type II topoisomerases, DNA gyrase and DNA topoisomerase IV. Type II topoisomerases are able to cut both strands of the DNA and pass one strand through this transient gap before resealing the DNA. Both topoisomerase enzymes are essential for bacterial growth, but they cannot complement one another [14, 15].

During transcription process or DNA synthesis, the double-stranded DNA is unzipped to allow either mRNA or a new DNA strand to be synthesized. This unzipping of the DNA causes topological stress upstream of the RNA polymerase or replication fork and induces the formation of positive supercoils that need to be removed. To relieve this stress and to remove the positive supercoils, a topoisomerase known as DNA gyrase makes double-stranded breaks in the DNA. With DNA gyrase, the cut DNA strand is passed through itself to propagate what is known as negative supercoiling. After DNA synthesis, the daughter chromosomes are unlinked by another topoisomerase, topoisomerase IV, in a process called decatenation. Both of these enzymes belong to the class of type II topoisomerases, use a double-strand-passage mode of action, are essential to bacteria and are the targets of the quinolones [16].

DNA gyrase is composed of two GyrA and two GyrB monomeric subunits, which are encoded by the *gyrA* and *gyrB* genes, respectively. Topoisomerase IV is composed of two ParC and two ParE (in *S. aureus* named GrlA and GrlB, respectively) subunits. The two subunits of this tetrameric enzyme are encoded by the *parC* and *parE* genes (referred to as *grlA* and *grlB* in *S. aureus*). GyrA is similar in structure to ParC, and GyrB is similar in structure to ParE. GyrA and GrlA are the subunits responsible for the physical cutting of the DNA, whereas GyrB and GrlB provide the energy for supercoiling of decatenation by the breakdown of ATP [17, 18].

The activity of quinolones stems primarily from the formation of ternary complexes between DNA and type II topoisomerases. Complex formation reversibly inhibits DNA synthesis and cell growth, which is probably responsible for the bacteriostatic action of quinolones. Their lethal action is thought to be a separate event from complex formation and probably relates to the release of free DNA ends from the DNA gyrase–quinolone complexes [19-23]. According to this model, four quinolone molecules are envisaged to bind to the single-stranded DNA regions opened up in the gyrase-DNA complex by covalent attachment of the two GyrA subunits to each complementary DNA strand. Two quinolones are envisaged to lie above the other pair of drug molecules, making hydrophobic interactions with each other through substituents on N-1 and C-8 positions. Binding to DNA strands is suggested to involve a hydrogen bonding domain on the drug comprising the C-3 carboxyl group and the ketone at C-4. Furthermore, Shen et al. postulated that the substituent on C-7 is involved in drug-enzyme interactions [24].

A key to understanding quinolone action in *S. aureus* is the observation that the drugs can act preferentially through gyrase, through topoisomerase IV, or through both targets in a manner dependent on quinolone structure. In general, several studies have highlighted substantial variations in the in-vitro inhibitory concentrations for DNA gyrase and topoisomerase IV, depending on both the bacterial species and the molecule being studied. These data, which are roughly consistent with MIC values and data obtained from analysis of resistant mutants, confirm that DNA gyrase is the preferred target of quinolones in Gram-negative bacteria. The situation is more complex in Gram-positive bacteria [25]. It was proposed that effects of quinolones are more severe on topoisomerase IV in Gram-positive bacteria (e.g., *S. aureus*) than its counterparts in Gram-negative bacteria (e.g., *E. coli*). This scenario is consistent with a species-specific phenomenon in which topoisomerase IV would be the primary target for all quinolones in *S. aureus*, for example. However, it was also recognized that sparfloxacin was a notable exception to the rule in that mutations in GrlA caused minimal loss of potency compared to other quinolones, such as pefloxacin, ciprofloxacin, levofloxacin and trovafloxacin against *S. aureus* [26, 27]. In addition, first-step mutations were found to map in *gyrA* when selected for reduced susceptibility to sparfloxacin, suggesting that DNA gyrase is the primary target for sparfloxacin in *S. aureus* [28, 29]. Furthermore, both DNA gyrase and topoisomerase IV appeared to act as dual targets for clinafloxacin and gemifloxacin [30, 31].

In a study by Nilius et al., trovafloxacin and ciprofloxacin showed 8- and 19-fold greater potencies, respectively, against *S. aureus* topoisomerase IV than against *S. aureus* DNA gyrase. In contrast, the interaction of ABT-492 with both *S. aureus* topoisomerases was nearly equivalent, with three-fold greater activity against DNA gyrase than against topoisomerase IV. By direct comparison, ABT-492 was nearly three-fold more active than trovafloxacin and was about six-fold more active than ciprofloxacin against *S. aureus* DNA gyrase [32]. In addition, the effective target inhibition data suggest that non-fluorinated quinolones have a dual-target mechanism of action against *S. aureus* and their potency against *S. aureus* could be attributable to partial inhibition of both the targets [33].

Concisely, quinolones can be classified into 3 categories (types I, II and III) based on their ability to inhibit purified *S. aureus* DNA gyrase and topoisomerase IV and also the effect of *grlA* and *gyrA* mutations on MIC [34, 35]. Most quinolones fall within the type I category (norfloxacin, enoxacin, fleroxacin, ciprofloxacin, lomefloxacin, trovafloxacin, grepafloxacin, ofloxacin, levofloxacin and gemifloxacin) because they are more influenced by topoisomerase IV inhibition than DNA gyrase inhibition. The type II category including sparfloxacin and nadifloxacin show a greater decrease in activity against a *gyrA* mutant strain than a *grlA* mutant strain – indicating a preference for DNA gyrase [34]. It should be noted however, that the results of some studies do not concur with this analysis for sparfloxacin [36, 37]. The type III quinolones include clinafloxacin, gatifloxacin, moxifloxacin and pazufloxacin, which have dual-targeting properties where either enzyme is inhibited at approximately the same extent and *gyrA* or *grlA* mutant strains produce only a minimal 2-fold increase in MIC [34]. Other data suggest that sitafloxacin [37] and garenoxacin [38] are also type III quinolones.

Chapter 3

QUINOLONE RESISTANCE IN *S. AUREUS*

Quinolone resistance in *S. aureus* emerged rapidly in several hospitals after the introduction of ciprofloxacin and strikingly so in MRSA, rising to prevalence of over 80% in some cases [39, 40]. The early finding with the older quinolone ciprofloxacin indicated in vitro activity against both MSSA and MRSA, although marginal with $MIC_{90}s$ of 0.5µg/mL [41]. Unfortunately, after few years almost all MRSA had developed resistance to quinolones. Blumberg et al. reported that prior to the introduction of ciprofloxacin in 1988, 159 MSSA and 131 MRSA were found to be susceptible to ciprofloxacin. One year after introduction, 79% of MRSA and 13.6% of MSSA were ciprofloxacin resistant [39]. This situation is also illustrated well by a study in Hong Kong where resistance to ciprofloxacin in MRSA had reached 82% in 1993 from a relatively low level of 9% in 1988 [42].

The decreased susceptibility of MRSA to quinolones is not the result of increased mutation rates; MSSA have been shown to have the same mutation rates as MRSA [43]. The early findings demonstrated that although initial cases of colonization or infection with quinolone-resistance MRSA can be related to quinolone use, the spread of MRSA is more likely due to hospital transmission of existing resistant MRSA. Thereby prevalence of MRSA over MSSA due to greater opportunities for transmission of MRSA between hospitalized patients, many of whom have received antibiotics. Because MRSA are commonly multidrug-resistant, selective advantage of these strains over MSSA strains may occur with a broad range of antibiotic exposures, not just quinolones [44].

There does appear to be a relation between quinolone resistance and methicillin resistance. It seems likely that an association exists between the action of quinolones on mec(A)-positive *S. aureus* and the increase in the proportion of CFU/mL expressing methicillin resistance. Growth in the presence of various quinolones increased the proportion of the mec(A)-positive heteroresistant *S. aureus* that developed methicillin resistance (MIC of oxacillin >128 µg/mL). The resultant MRSA strains also showed a 1.5- to 3-fold increase in quinolone MICs [43].

MECHANISMS OF QUINOLONE RESISTANCE IN *S. AUREUS*

The resistance of *S. aureus* to quinolones involves at least three different mechanisms, which are often combined in highly resistant organisms. One mechanism is the active efflux of the drugs by the NorA transporter [45]. The two other mechanisms result from modifications of the quinolone molecular targets of the bacterium, i.e., the DNA gyrase (*gyrA* mutants) [46] and/or the topoisomerase IV (*grlA* mutants) [47]. As with other bacterial species, resistance to quinolones in *S. aureus* develops in a step-wise fashion by the accumulation of mutations. Resistance occurs due to changes in aminoacids (Table 1), particularly those in certain regions of each enzyme subunit called the quinolone-resistance-determining-region (QRDR), which make the enzyme less sensitive to inhibition by quinolones. For some of the more common QRDR mutations it seems that the aminoacid changes reduce the affinity of the enzyme-DNA complex for the quinolone [48]. Level of resistance of a target enzyme can increase as additional aminoacids are altered by additional mutations [49].

Experimental studies have shown that mutations in *grlA* occur before *gyrA* with *S. aureus* selected on ciprofloxacin and these latter mutations are silent in the absence of additional mutations in *grlA* [50]. This is in general agreement with ciprofloxacin being a type I quinolone based on target preference. With sparfloxacin, the reverse is true with mutations in *gyrA* occurring before *grlA* [28] again in keeping with the target preference of a type II quinolone. The same classical quinolone resistance alteration in GrlA (Ser80 to Phe or Tyr and Glu84 to Lys) and GyrA (Ser84 to Leu or Lys and Glu88 to Lys or Val) have been found in laboratory mutants and clinical isolates [23, 51]. Other mutations such as Asp73 to Gly in GyrA have been fond clinically but are considerably less common [52].

Resistance mutations that alter the structure of the GyrB and ParE subunits are less common than those altering the structure of GyrA and ParC subunits. Single changes in ParC seem to be sufficient to cause clinical resistance to ciprofloxacin, but not necessarily to other more potent quinolones [51]. For more potent quinolones in staphylococci, single ParC mutations often generate reduced susceptibility without full resistance as defined in the clinical laboratory [51, 54, 55].

A mutation in the more sensitive enzyme results in an increase in the MIC of a quinolone, whereas a mutation in the less sensitive enzyme generally causes resistance only in the presence of resistance mutations in the primary target. A quinolone with similar affinities for both targets is little affected by a mutation in one of the enzymes, and concurrent mutations in both enzymes are necessary for resistance to develop [30, 56-58]. Thus, for dual-targeting quinolones with similar potency against both target enzymes, there may be little reduction in susceptibility with any single mutation in either topoisomerase IV or DNA gyrase [30].

Table 1. Common mutations in type II topoisomerases of *S. aureus* and MICs of ciprofloxacin against clinical isolates of *S. aureus*

Alteration of amino acid		MIC (μg/mL)	References
GyrA	GrlA		
–	–	0.25	[47]
–	Ser80→Phe/Tyr	2–16	[51], [52]
Ser84→Leu	–	16	[53]
Ser84→Leu	Ser80→Tyr	16	[51]
Ser84→Leu	Ser80→ Phe/Tyr	12.5–800	[52]
Ser84→Leu	Glu84→ Lys	100–800	[52]
Ser84→Leu	Ser80→ Phe/Tyr + Glu84Lys	100–>800	[52]
Glu88→Lys	Ser80→ Phe	8–32	[51]
Glu88→Lys	Ser80→ Phe + Glu84Val	32–256	[51]

Several studies pay attention to identify the structural features of quinolones and provide a rationale for dissociated propensity for development of resistance amongst *S. aureus* and other Gram-positive bacteria. Concerning on the C-6 position, the primary objective of the non-fluorinated quinolones design effort was achieved with a series of quinolones that are relatively unaffected by the serine mutation in DNA gyrase. However, the data reviewed by Roychoudhury et al. showed that those non-fluorinated quinolones also have highly desirable properties against the second target of quinolones, topoisomerase IV, especially in *S. aureus* [33, 59]. Hartman-Neumann et al. found that the non-fluorinated quinolone, garenoxacin able to select mutants with the known changes in GyrA and ParC [60].

Fukuda et al. demonstrated the importance of an 8-methoxy substituent by comparing gatifloxacin and AM-1147 (an 8-methoxy quinolone) and their respective 8-unsubstituted counterparts. They showed that the 8-methoxy derivatives to select mutants at a lower frequency than their 8-unsubstituted counterparts and to prefer DNA gyrase [61]. Furthermore, Ince and Hooper [62] underlined the importance of the 8-methoxy substituent by comparing gatifloxacin, its 8-desmethoxy derivative and ciprofloxacin. Gatifloxacin was found to be most active against mutants of *S. aureus*, and a novel substitution outside the QRDR of GrlA was detected: Lys23→Asn. Two novel mutations in *grlB* were also found: Pro25→His and Pro451→Gln. Allelic exchange experiments confirmed the role of the novel mutations in resistance, suggesting that the QRDR in GrlA should be expanded to include these mutations. Zhao et al. showed that 8-methoxy ciprofloxacin derivatives are more lethal than 8-bromo, 8-ethoxy, and 8-H derivatives for *S. aureus*, especially when topoisomerase IV is resistant. The methoxy group also increases lethality against wild-type cells when protein synthesis was inhibited [63].

ROLE OF EFFLUX PUMPS

The second resistance mechanism in *S. aureus* is mediated through active efflux by the over-expression of certain efflux pumps. The efflux pump system is a mechanism that allows immediate survival of bacteria in the presence of an antimicrobial agent by actively expelling that agent across the cell membrane, thereby reducing the intracellular concentrations to sub-lethal levels [21, 64].

Low level of resistance to quinolones can occur by over-expression of these efflux pumps, usually due to mutations that increase transcription of the structural gene for the pump either from effects on the gene promoter or from alterations in other regulatory proteins that affect pump expression. In the case of *S. aureus* including MRSA, low level of resistance to ciprofloxacin can result from the over expression of a pump called NorA [21]. Mutations causing this increase and low-level resistance have been seen in the *norA* gene promoter causing increased stability of *norA* mRNA and in other regulatory proteins [21]. Because NorA is a multi-drug efflux pump, its increased expression also reduces susceptibility to chloramphenicol and some dyes, such as ethidium bromide. NorA also contributes indirectly to other types of resistance because exposure of bacteria to sub-therapeutic drug concentrations promotes the selection and expression of higher-level adaptive resistance mechanisms, such at target mutations [65, 66].

Sulavik and Barg [67] examined mutants of clinical isolates of MRSA and MSSA with low-level resistance to ciprofloxacin. The comparison of generated first-step mutants and mutants with low-level ciprofloxacin resistance showed that in vitro selection did not alter the distribution into different classes of first-step mutants, with about 10% of the first-step mutants showing increased efflux.

The efflux pump action is dependent on the ability of quinolone molecule to bind to the bacterial efflux protein and be exported [68]. Norfloxacin, ciprofloxacin and ofloxacin seem to be affected more than sparfloxacin, gatifloxacin, sitafloxacin, gemifloxacin, moxifloxacin, trovafloxacin and garenoxacin, suggesting that these last quinolones may be poorer substrates for these pumps [21, 64, 69-71]. Thus, efflux-mediated resistance mechanisms seem to affect quinolone agents to different extents, depending on the physicochemical properties and structural characteristics of the individual quinolones. Previously, only hydrophilic character of the fluoroquinolones was thought to be an important factor for NorA-mediated transport. Aeschlimann et al. [70] examined the in vitro antibacterial activities and post-antibiotic effects of ciprofloxacin, norfloxacin and levofloxacin, in genetically related strains of *S. aureus*. They used the wild-type and two NorA-mediated resistant strains for susceptibility testing and determination of time-kill curves and post-antibiotic effect. It was found that the NorA inhibitors, reserpine and omeprazole dramatically improve the activities of the more hydrophilic quinolones (ciprofloxacin and norfloxacin). In addition, hydrophilic quinolones, such as norfloxacin and ciprofloxacin, appear more prone to efflux than more hydrophobic molecules like grepafloxacin and gatifloxacin in Gram-positive organisms [72, 73]. However, Takenouchi et al. demonstrated that the bulkiness of the C-7 substituent and the hydrophobicity and bulkiness of the C-8 substituent, not the molecular hydrophobicity, was correlated with the activity of quinolones [74]. Of the quinolones tested by Beyer et al. and Madras-Kelly et al., levofloxacin has the least bulky C-7 substituent (*N*-methylpiperazine), sparfloxacin a slightly larger derivative (bearing a 3,5-dimethylpiperazine) and moxifloxacin, the bulkiest C-7 substituent (diazabicyclo moiety). On the other hand, sparfloxacin is a rather hydrophobic drug, while levofloxacin and moxifloxacin are much more hydrophilic. Sparfloxacin and moxifloxacin were relatively unaffected by NorA-mediated efflux, whereas levofloxacin was highly affected [69, 75].

In order to find quinolones that are refractory to efflux-mediated resistance in Gram-positive bacteria, Kerns et al. discovered a novel series of piperazinyl-linked quinolone dimers, which exhibited potent antibacterial activity against drug resistant strains of *S. aureus*, including strains possessing resistance due to the NorA multidrug efflux pump and a mutation in the QRDR of topoisomerase IV. They suggested that a bulky C-7 side chain might account for the lack of effective dimer efflux through the *norA* pump [76, 77]. These data demonstrate that the bulk at C-7 appears to be the key structural characteristic for avoidance of efflux. Gemifloxacin, moxifloxacin, trovafloxacin, and garenoxacin are the currently available agents with the greatest bulk at C-7 position, and they appear least affected by efflux-mediated resistance [25, 64].

Chapter 4

GENERAL STRUCTURAL FEATURES (STRUCTURE-ACTIVITY RELATIONSHIPS)

Progressive efforts have been made to optimize the biological activity of the quinolones against Gram-positive bacteria including *S. aureus*. The minimum pharmacophore required for significant antibacterial activity consists of the 4-pyridone ring with a 3-carboxylic acid group (Figure 5). The nitrogen at N-1 is almost indispensable for activity and the carboxylic acid at position 3 and the carbonyl group at position 4 are considered critical for binding to cleaved or perturbed DNA. Therefore, the 3-carboxylate and 4-carbonyl groups are considered essential for antimicrobial activity [78]. Thus, the 1,4-dihydro-4-oxopyridine-3-carboxylic acid associated with a 5,6-fused aromatic ring is the common chemical feature of quinolones and 1-, 2-, 5-, 6-, 7-, and 8-positions are the major targets of chemical variation (Figure 5). The antibacterial activity of quinolones depends on the nature of peripheral substituents and their spatial arrangements [57]. The purpose of our brief discussion about structure-activity relationships (SAR) of quinolones is to demonstrate the changes or effects occurred due to modifications of different substituents in particular position of the pharmacophore, and influenced the antibacterial activity especially against *S. aureus*.

POSITION 1

Antibacterial activity is greatly influenced by the hydrophobic and steric bulk of N-1 substituent [79]. A cyclopropyl substituent is now considered as a particularly potent modification at this position; most of the successful compounds against *S. aureus* contain this substituent including ciprofloxacin, sparfloxacin, grepafloxacin, gatifloxacin, clinafloxacin, gemifloxacin, moxifloxacin, and garenoxacin. Sitafloxacin contains a fluorinated cyclopropyl group [*cis*-oriented (1*R*,2*S*)-2-fluorocyclopropyl] at N-1 [80, 81]. The addition of a 2,4-difluorophenyl group at position 1 [e.g. tosufloxacin and trovafloxacin] also improved activity against *S. aureus* as well as other Gram-positive bacteria [82]. It has been observed that in N-1 aryl substituted quinolones, the N-1 aryl ring should be oriented out of the plane of quinolone ring for its antibacterial activity [83]. Another distinct structure at this position is found in ofloxacin, levofloxacin, pazufloxacin and nadifloxacin (Figure 6) which have a fused ring between positions 1 and 8. For instance, ofloxacin has a tricyclic ring structure with a methyl group attached to the asymmetric C-3 position on the oxazine ring, thus connecting positions 1 and 8 with a fused ring. Although this has been a useful alternative to the cyclopropyl substituent, it has been also shown that the *S*- isomer (levofloxacin) is 2-4 times more active than ofloxacin against both methicillin-susceptible and resistant isolate of quinolone-susceptible staphylococci [84].

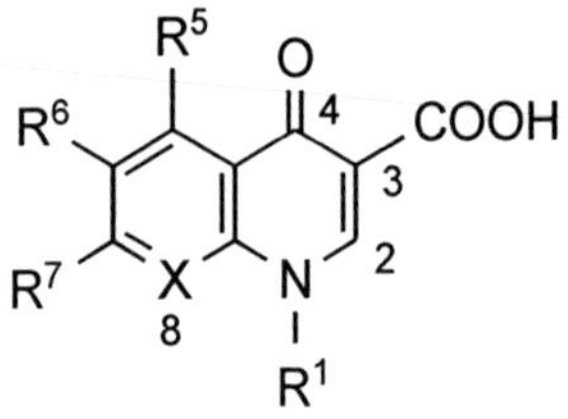

Figure 5. 4-Quinolone-3-carboxylic acid; common chemical feature of quinolone antibacterials and 1-, 5-, 6-, 7-, and 8-positions as the major targets of chemical variation.

POSITION 2

In general, no analog modified at C-2 position remains on the market bearing other than a CH substituent. Very little is known about the SAR of quinolone having substituents at C-2 position; as loss of bioactivity has been found with methyl, hydroxyl or methylthio substituents. Moreover, significant decrease in activity was found with 2-aza-quinolones derivatives of norfloxacin. However a ring between C-1 and C-2 position was shown to have biological activity. The C-2 position is left unsubstituted because of its proximity to the enzyme binding site [7, 9, 10, 85]. In addition, It is seems that the COOH at C-3 needs to be coplanar with the C-4 carbonyl group so that it can effectively hydrogen bond with DNA bases made available by strand separation catalyzed by type II topoisomerases. The steric deficit that substitution at C-2 presumably causes can be partly overcome by rigidifying a C-2 substituent in a ring connecting it with N-1, especially when the group attached to C-2 can accept a hydrogen bond and form a virtual ring with the C-3 substituent [87]. In agreement with this idea is the outstanding potency seen with formation of a thiazolone ring fused to the carbonyl ring [88]. With thiazolone ring formation, the NH moiety is strongly acidified by resonance stemming from the aromaticity of the ring when enolization takes place. This produces a coplanar carboxyl surrogate and leads to an enhancement of antibacterial activity.

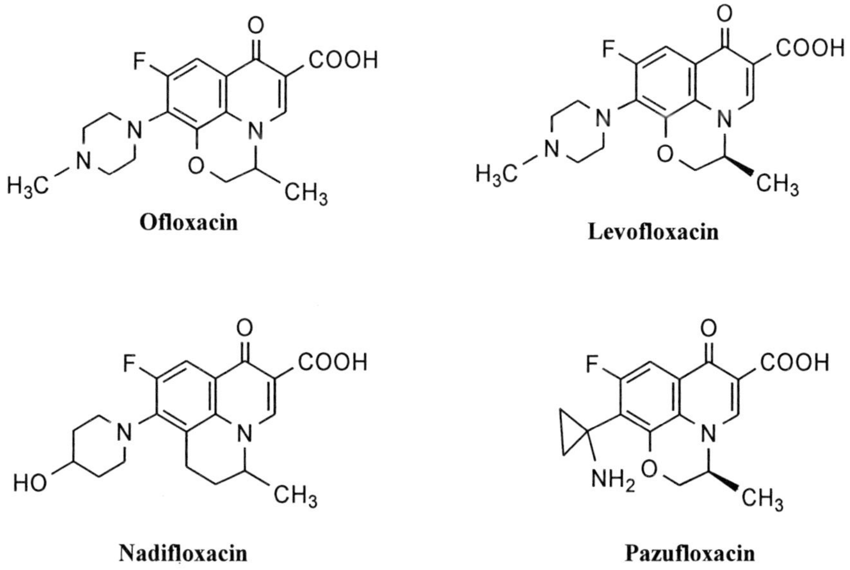

Figure 6. Quinolones with a fused ring between positions 1 and 8.

POSITION 5

Introduction of some substituents such as halogen, nitro, amino, hydroxy and alkyl groups at C-5 were initially thought to reduce antibacterial activity of the quinolones. Subsequently, it has been shown that introduction of bulky substituents, halogen and methoxy at this position reduce activity markedly; this is probably a consequence of interference with the active binding site at positions 3 and 4 [10, 78]. However, modestly sized substituents, such as an amino (as in sparfloxacin) or a methyl group (as in grepafloxacin) can markedly increase in vitro activity against *S. aureus* [10, 89]. Thus, substitutions at this position are thought to contribute to potency against *S. aureus* and other Gram-positive microorganisms [90-92]. It is also notable that the influence of C-5 substituent markedly depends on the substitution pattern at N-1, C-7 and C-8. A large number of current quinolones including ciprofloxacin, ofloxacin, levofloxacin, gatifloxacin, clinafloxacin, sitafloxacin, gemifloxacin, moxifloxacin, trovafloxacin and garenoxacin have only hydrogen at this position.

POSITION 6

Early investigations on the C-6 position were generally limited to the terminus of a CH (nalidixic acid), methylenedioxy bridge to C-7 (oxolinic acid and cinoxacin), and N (pipemidic acid and piromidic acid) (Figure 1). These early compounds had moderate activity against Gram-negatives and no activity against Gram-positives. The addition of fluorine substituent at C-6 markedly improved antimicrobial activity [93, 94]. Flumequine was the first compound to be developed with a fluoro- group at C-6, and gave the first indications that modifications of the basic chemical structure could improve Gram-positive activity [5]. The influence of fluorine at this position is essential for high activity as evidenced by its enhanced gyrase inhibition and cell penetration which has become the basis for generic name fluoroquinolones [57, 94]. After introduction of norfloxacin and attractive properties of 6-fluoro- substituent, a large number of 6-fluoroquinolones were marketed. Subsequently, it has been shown in a series of quinolones that the impact of 6-fluoro- substituent is diminished when the molecule contains spatial arrangements with other helpful substituents. For example, studies with garenoxacin (a 6-desfluorinated-8-difluoromethoxyquinolone, Figure 4) and its 6-fluoro-analog revealed no difference in potency against *S. aureus* whether the C-6 substituent is fluorine or hydrogen. The objective of the 6-desfluoro-quinolones optimization effort was to design compounds with potent activity against important Gram-positive pathogens *S. aureus* and *S. pneumoniae* and adequate activity against MRSA strains, while maintaining broad-spectrum antibacterial activity [33, 59, 95, 96]. During the initial investigation of the 6-desfluoro-quinolones, it was noticed that removal of the fluorine at the position C-6 systematically lowered the level of genotoxicity, based on a comparison of the non-fluorinated quinolones and the corresponding 6-fluoroquinolones [33]. It then became clear that the 6-desfluoro-quinolones series was a useful platform for reoptimization of the quinolone backbone toward improved potency against *S. aureus* and *S.*

pneumoniae. The 6-desfluoro-quinolones, for example garenoxacin, show greater potency than the newer fluoroquinolone moxifloxacin against both sensitive and resistant strains of these organisms [56].

Moreover, recent studies revealed that good antibacterial activity was retained by replacing the C-6 fluorine atom with an amino- group and reoptimizing the other substituents. For the 6-amino compounds as well, the overall potency is highly dependent on substituents at positions C-7 and C-8. Thus, two 6-aminoquinolones I and II (illustrated in Figure 7), displayed the best activity against Gram-positive bacteria. Further structural modification of 6-aminoquinolones led to the discovery of 6-amino-8-methyl-quinolone (III), which is about 50-times more potent than ciprofloxacin against *S. aureus* [97-99]. The fluorine at C-6 is also replaced by nitro- group to get highly potent nitroquinolones [100]. They are potent inhibitor of *Streptococcus* and *Staphylococcus*. Other C-6 substituents that have been investigated, with less satisfactory results, are Cl, Br, acetyl, CN, and methyl groups.

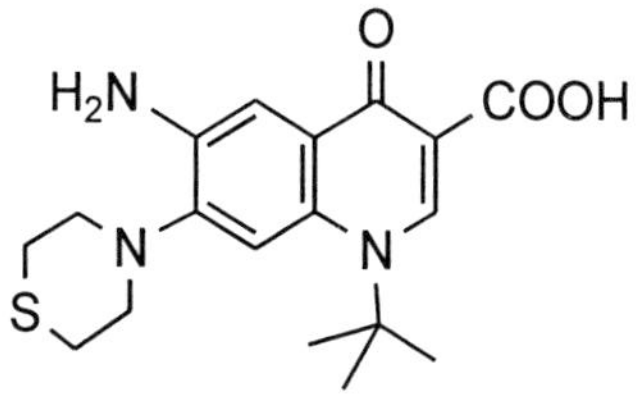

6-Aminoquinolone I

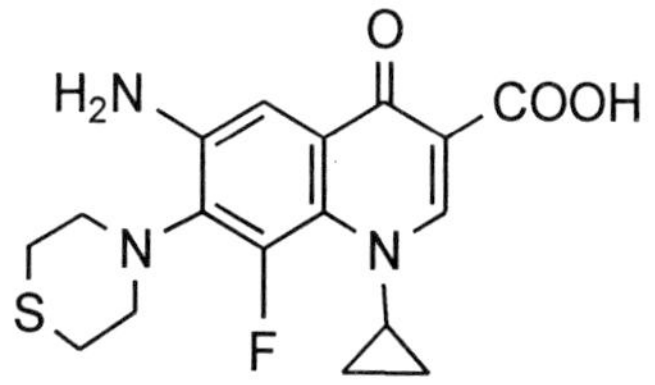

6-Aminoquinolone II

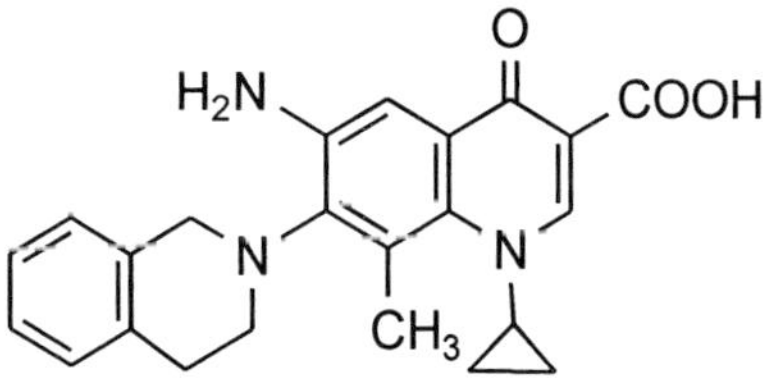

6-Aminoquinolone III

Figure 7. 6-Aminoquinolones.

POSITION 7

Many thousands of analogs have been prepared employing various substituents at C-7 position, leading to the conclusion that a cyclic amino-moiety is usually best. Furthermore, this 7-cyclic amine moiety of quinolones possesses enough structural flexibility to allow product optimization. The substituents at position C-7 are associated with a number of key attributes, such as antibacterial spectrum, bioavailability and adverse effects. The nature of the C-7 substituent also strongly affects the target preferences (DNA gyrase and/or topoisomerase IV) of quinolones. A five- or six-membered cycloamino moiety (such as, pyrrolidine or piperazine rings, respectively) is the most commonly found substitution at C-7 position. After the discovery of norfloxacin the piperazinyl moiety or its *N*-methyl analogue became a standard feature among quinolones. Unsubstituted piperazine rings [for example in norfloxacin, enoxacin or ciprofloxacin] confer potency against Gram-negative bacteria, while the addition of methyl groups can improve both oral absorption and anti-staphylococcal activity (generally, anti-Gram-positive activity) [5, 29, 101]. Pefloxacin, fleroxacin, ofloxacin and levofloxacin have a *N*-methyl piperazine, lomefloxacin, grepafloxacin and gatifloxacin have a 3-methyl piperazine, and sparfloxacin has a 3,5-dimethyl piperazine. These substituted piperazine-containing compounds have greater activity against Gram-positives, and are believed to have enhanced penetration into the bacterial cell [10, 102].

Pyrrolidinyl substituents with a pendant amine are also common and suitable substitution at position 7, and are associated with enhanced activity against Gram-positives. Clinafloxacin and tosufloxacin both have a 3-amino pyrrolidine substituent and sitafloxacin has a spiro-aminopyrrolidine group at C-7 [78, 80]. Similar to piperazine ring, introduction of methyl groups on the pyrrolidine ring also enhances activity against Gram-positive pathogens and helps to overcome some physicochemical and pharmacokinetics disadvantages [86, 101].

The introduction of bicyclic amino- groups onto position 7 has resulted in agents (moxifloxacin and trovafloxacin) with significant anti-Gram-positive activity and marked lipophilicity [82, 103]. Moxifloxacin has a diazabicyclic ring, while trovafloxacin has an amino-azabicyclic ring [82, 103, 104].

A number of quinolones (e.g., garenoxacin) have a certain aryl moiety instead of cyclic amine at C-7. Garenoxacin has (isoindolin-5-yl) moiety at C-7 with outstanding Gram-positive coverage, especially against *S. aureus* and *S. pneumoniae* [56, 105].

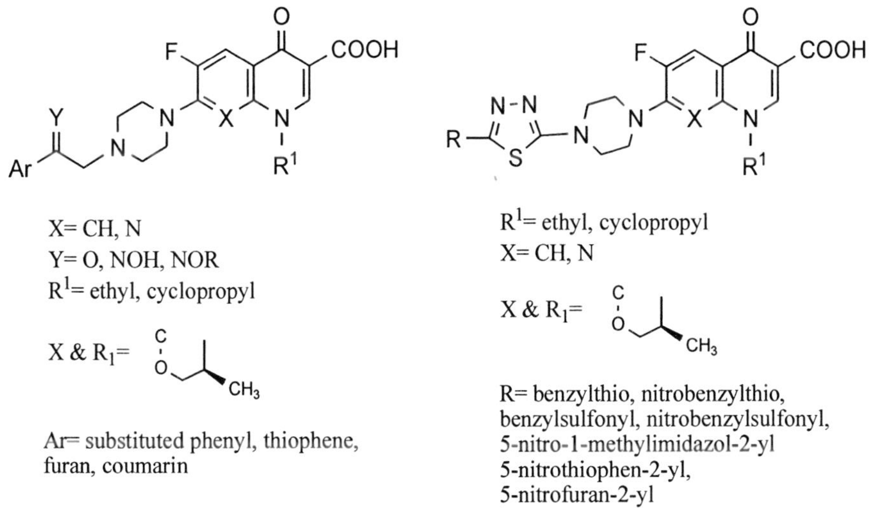

X= CH, N
Y= O, NOH, NOR
R¹= ethyl, cyclopropyl

X & R₁= $\overset{\text{C}}{\underset{\text{O}}{\cdot}}\diagdown\underset{\text{CH}_3}{\diagdown}$

Ar= substituted phenyl, thiophene, furan, coumarin

R¹= ethyl, cyclopropyl
X= CH, N

X & R₁= $\overset{\text{C}}{\underset{\text{O}}{\cdot}}\diagdown\underset{\text{CH}_3}{\diagdown}$

R= benzylthio, nitrobenzylthio, benzylsulfonyl, nitrobenzylsulfonyl, 5-nitro-1-methylimidazol-2-yl
5-nitrothiophen-2-yl,
5-nitrofuran-2-yl

Figure 8. *N*-substituted piperazinyl quinolones.

The pyrrolidinyl or piperazinyl moiety attached to C-7 position of quinolones possesses enough structural flexibility to allow modification and product optimization. In addition, a position on the quinolone molecule, where substitutions of bulky groups are permitted, is C-7 position [24, 106]. In addition, it has been proposed that for *S. aureus* and other Gram-positive organisms, increasing molecular mass and bulkiness of substituent at C-7 position is not a barrier to penetration [107]. Staphylococci do not possess an outer membrane, and so lack outer membrane proteins and lipopolysaccharide. Accumulation of quinolone drugs by *S. aureus* is thought to take place by simple diffusion across the cytoplasmic membrane [108, 109]. Accordingly, comprehensive structure-activity relationship studies on moieties attached to the C-7 cyclic amine of quinolones have been explored to identify the optimum structural requirements for designing the new anti-staphylococcal agents. We described different series of *N*-substituted piperazinyl quinolones containing certain bulky substituent connected to the piperazine unit of 7-piperazinyl quinolones (Figure 8). Some of these derivatives exhibit high activity against staphylococci more potent than their parent *N*-piperazinyl quinolones [106, 110-123].

Because of high tolerance for structural variation at the C-7 position of the quinolone ring, several quinolone-containing hybrids via C-7 connection, including quinolone-nitrothiophene, quinolone-nitrofuran and quinolone-nitroimidazole hybrids have been synthesized and evaluated [119-123]. The comparison of MIC values of these hybrids and their parent quinolones revealed that some of them have a more potent antibacterial activity against *S. aureus* (Figure 9).

In order to find new C-7 modified quinolones, Kerns et al. described symmetric and asymmetric C-7-piperazinyl-linked dimers of quinolones (Figure 10) possessing potent antibacterial activity against drug-resistant strains of *S. aureus* [76, 77].

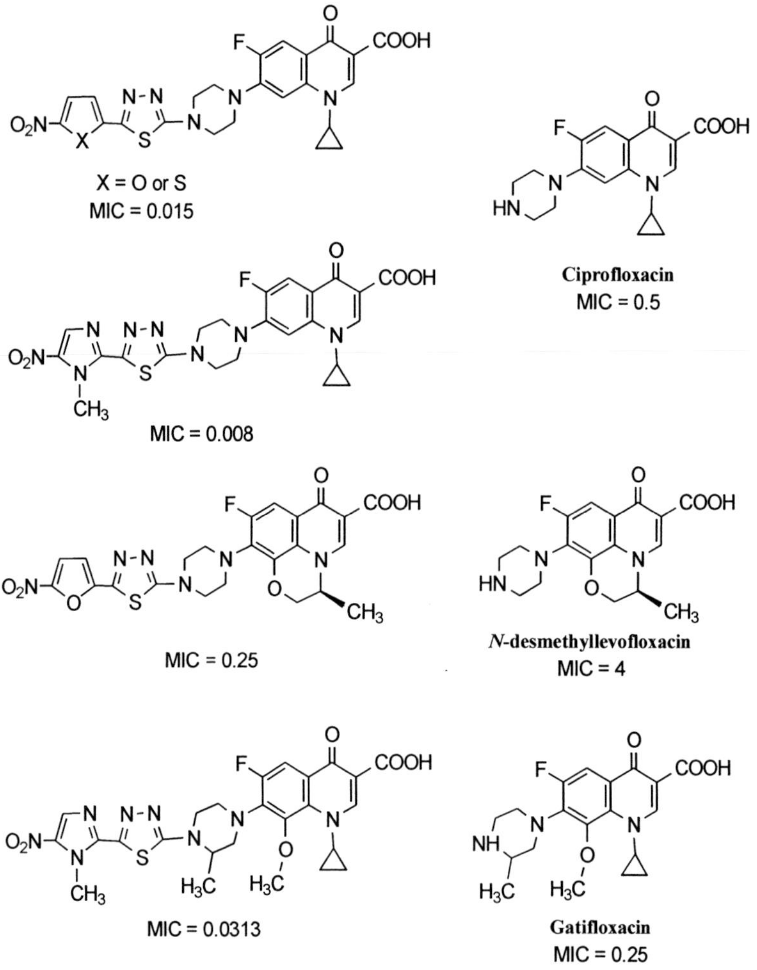

Figure 9. Quinolone-nitrofuran, quinolone-nitrothiophene and quinolone-nitroimidazole hybrids with more potent antibacterial activity against *S. aureus* relative to their parent quinolones (compared by MICs, µg/mL).

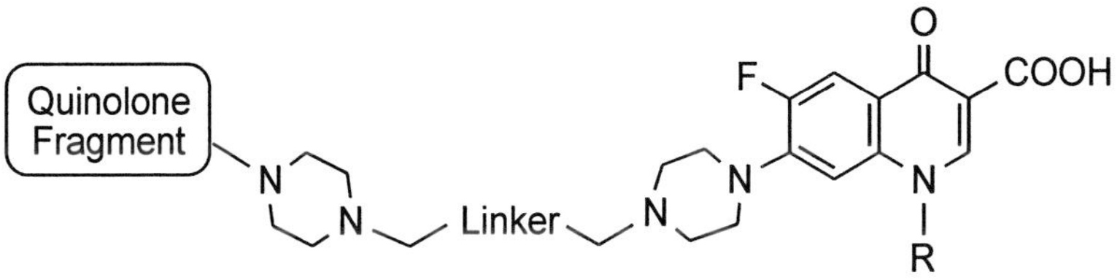

Figure 10. C-7-piperazinyl-linked dimers of quinolones, possessing potent antibacterial activity against drug-resistant strains of *S. aureus*.

POSITION 8

Substituents at C-8 position play an important role in altering oral pharmacokinetics, broadening the spectrum of activity and reducing the selection of mutants [63, 98, 124, 125]. Although, a number of quinolones such as grepafloxacin have no substituent at the position 8, but have a good activity against *S. aureus* and other Gram-positive pathogens. In addition, some naphthyridines (tosufloxacin, gemifloxacin, and trovafloxacin) in which C-8 of quinolone is replaced by a nitrogen atom, have excellent activity against *S. aureus*.

Halogen substituents, as well as a methyl or methoxy also increase the in vitro activity against *S. aureus*, even in those bacteria resistant to older fluoroquinolones. Early work showed that the 8-fluoro group (as in lomefloxacin and sparfloxacin) and 8-chloro group (as in clinafloxacin and sitafloxacin) were the most favorable substituents in terms of antibacterial activity; however, quinolones containing C-8 halogens tend to exhibit phototoxicity and other unacceptable side-effects. In contrast, newer quinolones such as gatifloxacin and moxifloxacin, with a methoxy group at the C-8 position exhibit enhanced activity against Gram-positives but does not seen to carry any risk of phototoxicity and cytotoxicity [11, 12, 85, 86, 103, 124, 125].

Lu et al. examined bacteriostatic and bactericidal activities of 12 fluoroquinolones against a gyrase-topoisomerase IV double mutant of *S. aureus*. They indicated that C-8 halogen and C-8 methoxy groups enhanced activity against *S. aureus* and the MIC_{99} was reduced seven- to eightfold for the *S. aureus* mutant by C-8 bromine, chlorine, and methoxy groups. Also, they showed that mutant *S. aureus* was affected more than the wild-type by the addition of a C-8 substituent. C-8 halogen and methoxy groups also improved the ability to kill staphylococci [63, 126].

Chapter 5

NEWER QUINOLONES WITH POTENTIAL ANTI-MRSA ACTIVITY

Several newly developed quinolones and related compounds including 6-fluoroquinolones (Figure 11), 6-fluoronaphthyridones (Figure 12), 6-desfluoroquinolones (Figure 13), quinazolinediones and 4-quinolizinones (2-pyridones) (Figure 14) exhibit enhanced activity against *S. aureus* and have proved useful against MRSA. The structures and activities of these categories of quinolones will be discussed in this section.

6-FLUOROQUINOLONES

DK-507k

DK-507k, 7-[(7*S*)-7-amino-5-azaspiro [2.4]heptan-5-yl]-6-fluoro-1-[(1*R*,2*S*)-2-fluoro-1-cyclopropyl]-1,4-dihydro-8-methoxy-4-oxo-3-quinolinecarboxylic acid hydrochloride hydrate (Daiichi, Japan) is an 8-methoxyquinolone bearing a modified 3-aminopyrrolidine.

DK-507k

It is structurally related to sitafloxacin. DK-507k possesses an 8-methoxy group instead of 8-chloro substituent of sitafloxacin. Chlorine substituent, as well as a methoxy at C-8 position increases the in vitro activity against *S. aureus*, even in those bacteria resistant to older fluoroquinolones. Several studies showed that the 8-chloro group (as in clinafloxacin and sitafloxacin) was the most favorable substituents in terms of antibacterial activity [12, 103]; however, quinolones containing C-8 halogens tend to exhibit phototoxicity and other unacceptable side-effects. In contrast, newer quinolones such as DK-507k, with a methoxy group at the C-8 position presumably exhibit enhanced activity against Gram-positives and lesser risk of phototoxicity.

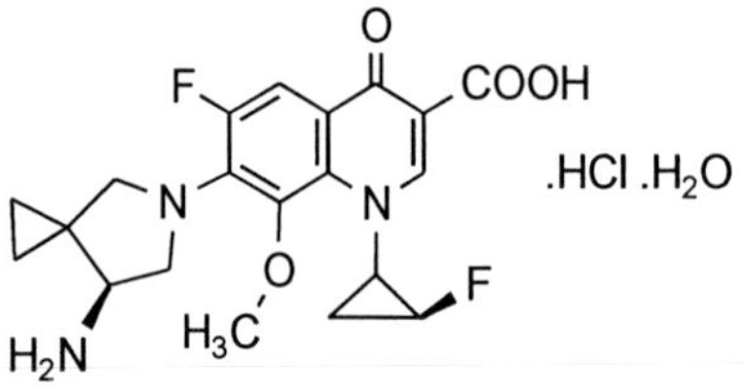

DK-507k

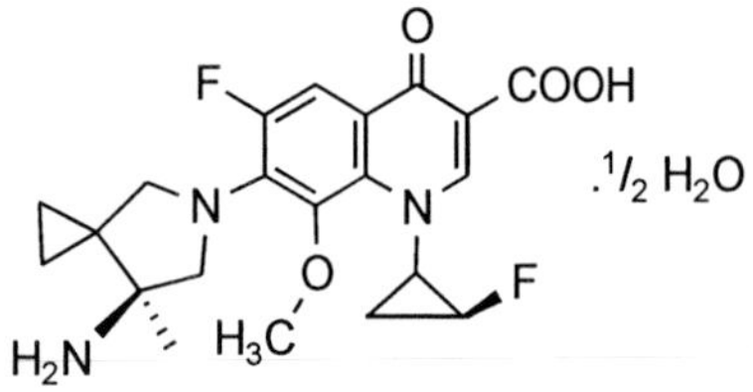

DC-159a

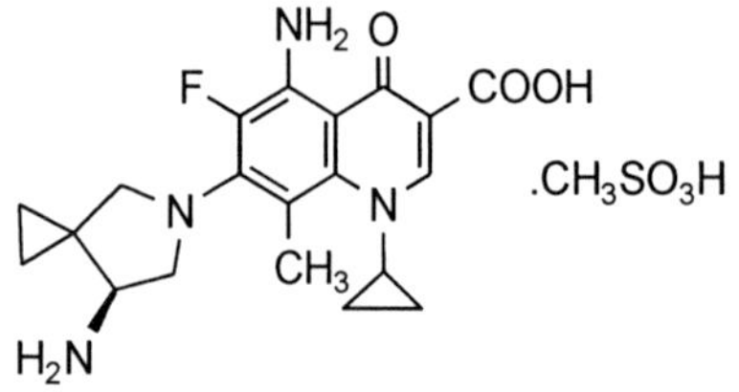

Olamufloxacin (HSR-903)

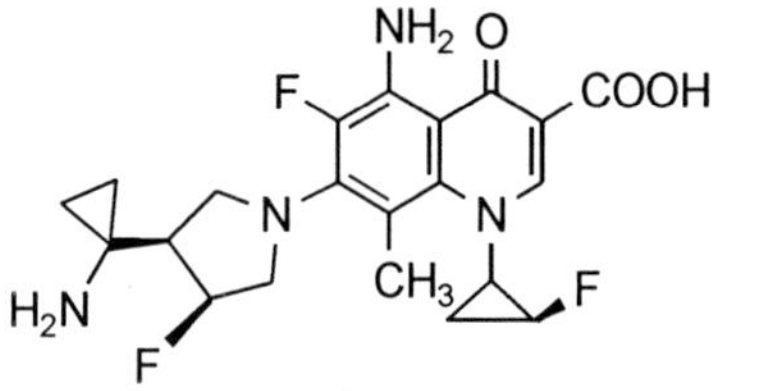

DQ-113

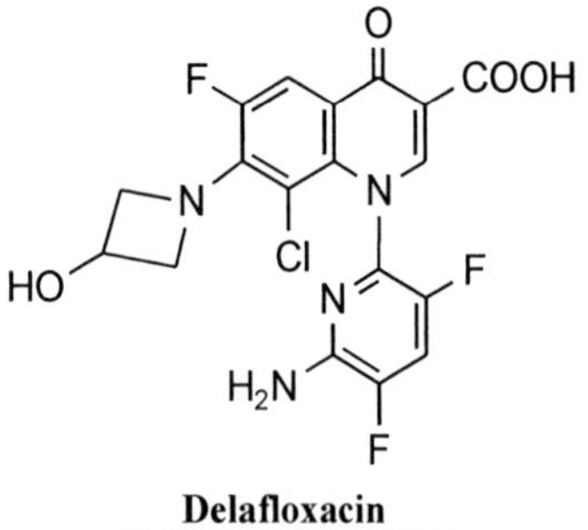

Delafloxacin
(ABT-492, RX-3341)

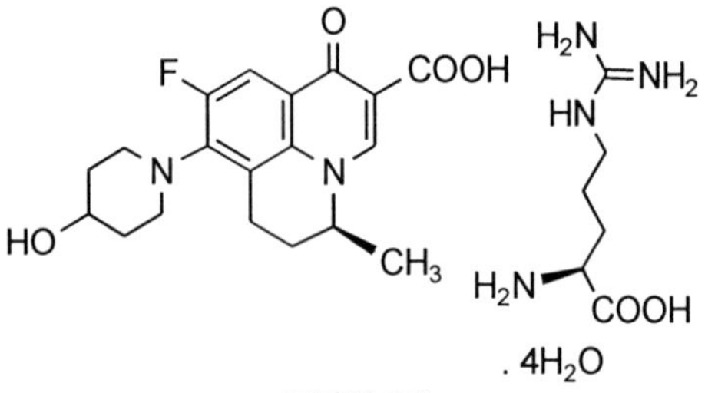

WCK 771

Figure 11. Newer 6-fluoroquinolones possessing potent anti-MRSA activity.

Gemifloxacin

DW286

Zabofloxacin
(DW-224a)

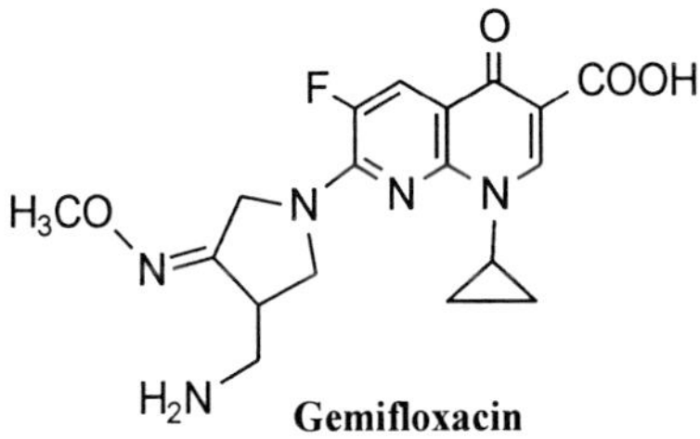

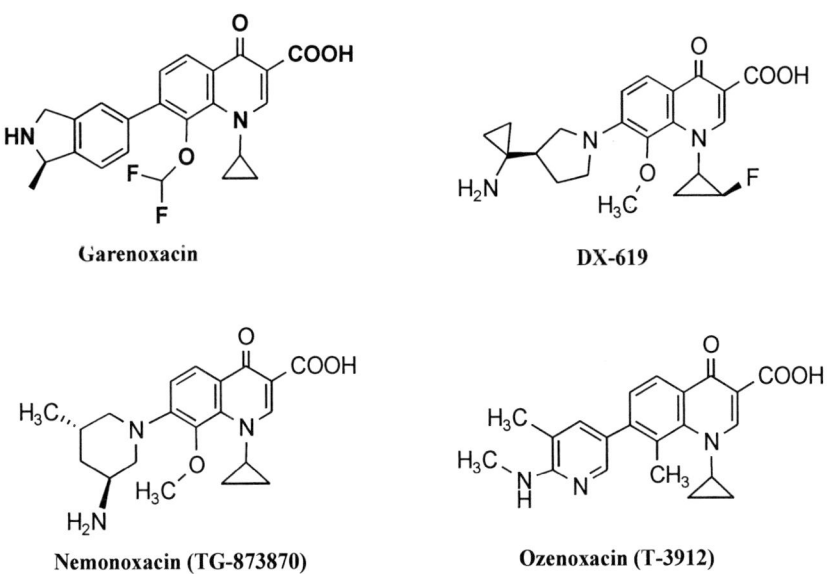

Figure 12. Newer 6-fluoronaphthyridones exhibiting potent anti-MRSA activity.

Garenoxacin

DX-619

Nemonoxacin (TG-873870)

Ozenoxacin (T-3912)

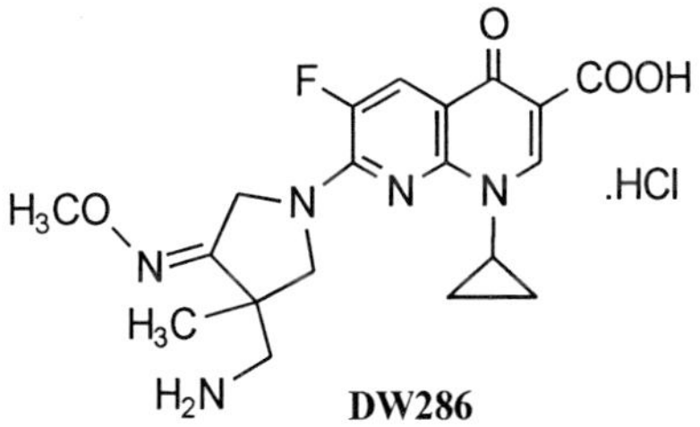

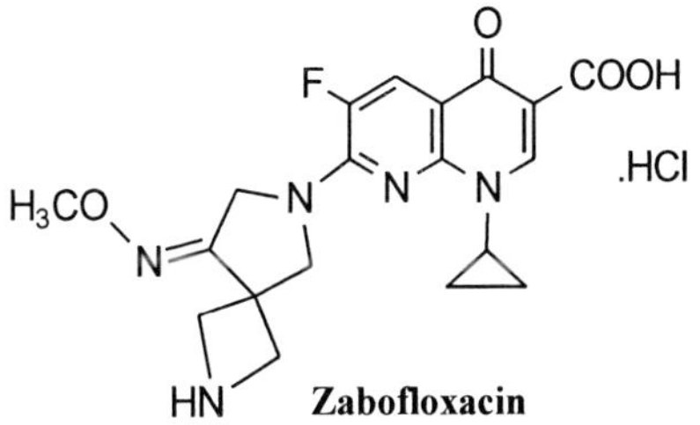

Figure 13. Newer 6-desfluoroquinolones having potent anti-MRSA activity.

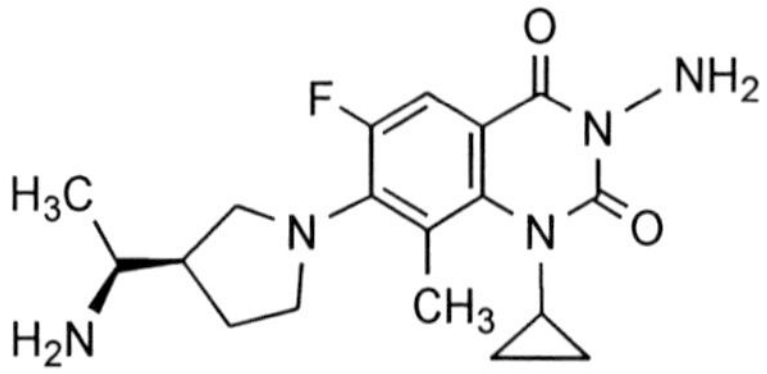

PD 0305970

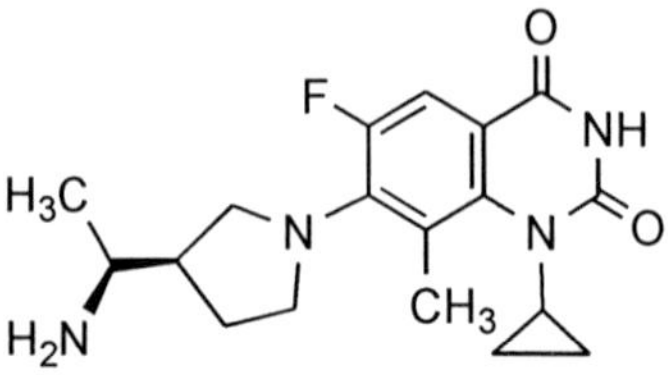

PD 0326448

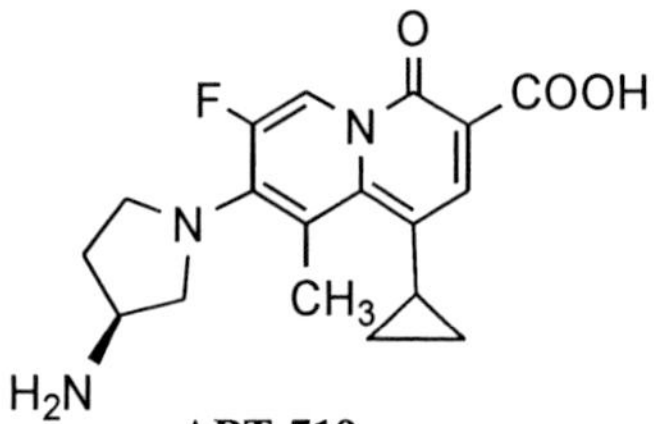

ABT-719

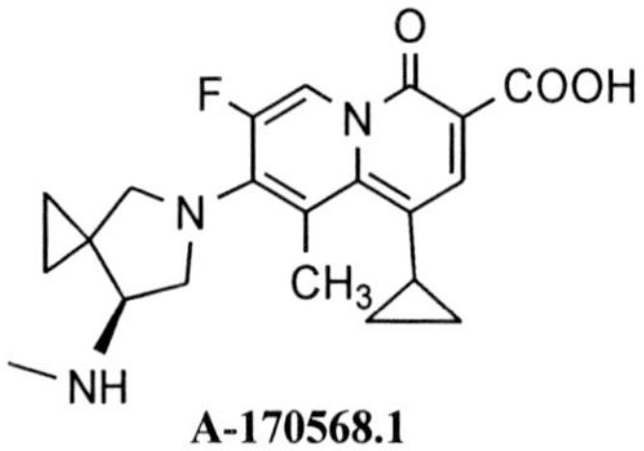

A-170568.1

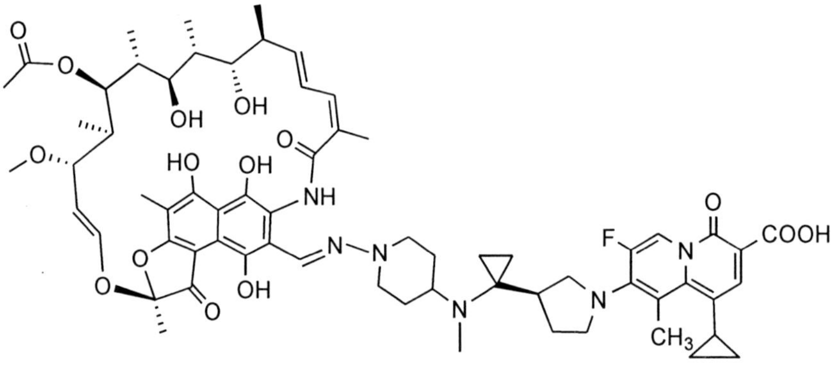

CBR-2092

Figure 14. Quinolone related compounds with potential anti-MRSA activity.

DK-507k had a broad antibacterial spectrum against both Gram-positive and Gram-negative bacteria [127]. Otani et al. compared the antibacterial activities of DK-507k with those of other quinolones including ciprofloxacin, gatifloxacin, levofloxacin, moxifloxacin, sitafloxacin, and garenoxacin. Against MSSA strains, DK-507k showed activities comparable to those of sitafloxacin and moxifloxacin, twofold lower than those of garenoxacin, and two- to fourfold higher than those of the other quinolones tested at the $MIC_{90}s$ (Table 2).

Table 2. Comparative MICs (µg/mL) of DK-507k and other quinolones against 25 strains of MSSA

Compound	X	R^1	R^6	R^7	Range	MIC_{50}	MIC_{90}
DK-507k	C–OCH₃	cyclopropyl-F	F	aminomethyl-spiro-azetidine (H₂N)	0.03–0.25	0.06	0.12
Sitafloxacin	C–Cl	cyclopropyl-F	F	aminomethyl-spiro-azetidine (H₂N)	0.008–0.12	0.03	0.12
Ciprofloxacin	C–H	cyclopropyl	F	piperazinyl (HN)	0.25–4	0.5	1
Gatifloxacin	C–OCH₃	cyclopropyl	F	3-methylpiperazinyl (HN, H₃C)	0.06–0.5	0.12	0.25
Moxifloxacin	C–OCH₃	cyclopropyl	F	octahydropyrrolo (HN, H)	0.06–0.25	0.06	0.12
Garenoxacin	C–OCHF₂	cyclopropyl	H	isoindoline (HN)	≤0.004–0.12	0.03	0.06
Levofloxacin	C–O–CH₃ (fused)		F	4-methylpiperazinyl (H₃CN)	0.12–1	0.25	0.5

Against ofloxacin-susceptible MRSA strains, DK-507k showed activities comparable to those of sitafloxacin; 2-fold lower than those of garenoxacin; and 2- to 16-fold higher than those of levofloxacin, ciprofloxacin, gatifloxacin, and moxifloxacin (Table 3). Furthermore, against ofloxacin-resistant MRSA strains the activities of DK-507k and sitafloxacin were the highest among those of the quinolones tested [128].

The efficacy of DK-507k in comparison with gatifloxacin, moxifloxacin and levofloxacin has been also evaluated against septicemia models with MSSA and MRSA in mice (Table 4). DK-507k was also the most active quinolone against MSSA and MRSA (ED_{50} = 1.07–9.23 mg/kg) [128].

Pharmacokinetics studies with radiolabeled DK-507k in rats and monkeys showed it to be rapidly absorbed after oral administration and to be excreted predominantly by the biliary route [129]. Urinary excretion in rats and monkeys was approximately 19.4 and 41.0% of the administered dose, respectively. Plasma protein binding in rats, monkeys and humans was about 40%.

DC-159a

DC-159a, (+)-7-[(7S)-7-amino-7-methyl-5-azaspiro [2.4]heptan-5-yl]-6-fluoro-1-[(1R,2S)-2-fluoro-1-cyclopropyl]-1,4- dihydro-8-methoxy-4-oxo-3-quinolinecarboxylic acid hemihydrate is a novel orally administered fluoroquinolone developed by Daiichi Pharmaceutical Co., Ltd. (Tokyo, Japan).

DC-159a

The distinctive substituents of DC-159a are a (1R,2S)-2-fluoro-1-cyclopropyl at N-1 position, a 7-[(7S)-7-amino-7-methyl-5-azaspiro [2.4]heptan-5-yl]- residue and an 8-methoxy group. The 2-fluoro-1-cyclopropyl group is also found in the structures of other new quinolones including sitafloxacin, DQ-113, DK-507k, and DX-619. The 7-amino-5-azaspiro [2.4]heptan-5-yl-residue is a new functionality for the C-7 position of quinolones which is found in sitafloxacin, DK-507k, A-170568.1 and HSR-903. The 8-methoxy group is also found in earlier quinolones such as gatifloxacin and moxifloxacin, and in new investigational quinolones for example DK-507k, DX-619, WCK 1152, WCK 1153 and nemonoxacin (TG-873870). Indeed, DC-159a is a methyl analog of DK-507k.

DC-159a is an 8-methoxy fluoroquinolone that possesses a broad spectrum of antibacterial activity, with extended activity against Gram-positive pathogens, especially streptococci and staphylococci from patients with community-acquired infections.

The antibacterial activities of DC-159a and the comparator drugs ciprofloxacin, moxifloxacin, garenoxacin, gemifloxacin and levofloxacin were evaluated by Hoshino et al. against Gram-positive and Gram -negative bacteria [130]. The MIC_{90} of DC-159a against MSSA and quinolone-susceptible MRSA was 0.06 µg/mL (Table 5). Against MSSA, DC-159a was 32-fold more active than ciprofloxacin, 8-fold more active than levofloxacin, 2-fold more active than moxifloxacin and gemifloxacin, and 2-fold less active than garenoxacin. Against quinolone-resistant MRSA, DC-159a (MIC_{90} = 8 µg/mL) showed the highest activity compared with the activities of the other quinolones tested. The MIC_{90} of DC-159a against methicillin-susceptible coagulase-negative staphylococci (MSCNS) and methicillin-resistant coagulase-negative staphylococci (MRCNS) was 0.5 µg/mL. Against MRCNS, DC-159a showed the highest activity compared with the activities of the other quinolones tested (Table 6).

In a separate study, Jones et al. evaluated the activity of DC-159a against clinical isolates of staphylococci (Table 7). DC-159a exhibited potent activity against MSSA strains, with MIC_{50}s and MIC_{90}s of 0.03 µg/mL [131]. This level of activity was 16-fold greater than that of ciprofloxacin but 2-fold less than that of gemifloxacin (MIC_{90} = 0.015 µg/mL). MRSA strains derived from patients with nosocomial infections showed the greatest susceptibility to DC-159a (MIC_{90} = 2 µg/mL). However, the range of DC-159a MICs was 32- to 64-fold greater than that for MSSA strains. The activity of DC-159a against 30 well-characterized isolates from patients with CA-MRSA infection was tested, and the MIC_{50} and MIC_{90} results for these strains were identical to those for the MSSA strains. Overall, among the five groups of staphylococci tested, DC-159a at ≤0.5 µg/mL inhibited all isolates in four of the groups (the exception was the nosocomial MRSA group). DC-159a showed a potency most similar to that of gemifloxacin (MIC_{90} range, 0.016 to 2 µg/mL) and markedly greater than the potencies of the other fluoroquinolones tested.

Table 3. Comparative MICs (μg/mL) of DK-507k and other quinolones against MRSA

Compound	X	R¹	R⁶	R⁷	MRSA (24), ofloxacin susceptible			MRSA (25), ofloxacin resistant		
					Range	MIC_{50}	MIC_{90}	Range	MIC_{50}	MIC_{90}
DK-507k	C-OCH₃	fluorocyclopropyl	F	aminospiro-pyrrolidine	0.015–0.06	0.03	0.06	0.25–32	1	4
Sitafloxacin	C-Cl	fluorocyclopropyl	F	aminospiro-pyrrolidine	0.008–0.06	0.03	0.06	0.25–32	1	4
Ciprofloxacin	C-H	cyclopropyl	F	piperazine	0.25–2	0.5	1	16–>128	128	>128
Gatifloxacin	C-OCH₃	cyclopropyl	F	methylpiperazine	0.06–0.25	0.12	0.25	2–>128	8	16

Table 3. Continued

Compound	X	R^1	R^6	R^7	Organism (number)					
					MRSA (24), ofloxacin susceptible			MRSA (25), ofloxacin resistant		
					Range	MIC_{50}	MIC_{90}	Range	MIC_{50}	MIC_{90}
Moxifloxacin	C-OCH$_3$	cyclopropyl	F	(bicyclic amine)	0.03–0.25	0.12	0.12	1–64	4	8
Garenoxacin	C-OCHF$_2$	cyclopropyl	H	(methylisoindoline)	0.008–0.06	0.03	0.03	0.25–64	2	8
Levofloxacin	C-O- (methyl oxazine)		F	H$_3$CN-piperazinyl	0.12–1	0.25	0.5	4->128	16	64

Table 4. Efficacy of DK-507k in comparison with gatifloxacin, moxifloxacin and levofloxacin on septicemia models in mice

Compound	Organism					
	S. aureus 037114 (MSSA)			S. aureus 037004 (MRSA)		
	MIC (µg/mL)	ED_{50} (mg/kg)	95% confidence interval (mg/kg)	MIC (µg/mL)	ED_{50} (mg/kg)	95% confidence interval (mg/kg)
DK-507k	0.06	1.07	0.74–1.45	1	9.23	5.14–16.94
Gatifloxacin	0.12	4.66	3.84–5.43	4	67.36	46.01–142.96
Moxifloxacin	0.06	5.48	3.34–9.82	2	56.70	45.39–73.08
Levofloxacin	0.25	16.20	12.02–19.57	16	>100.00	

Table 5. Chemical structures and comparative MICs (μg/mL) of DC-159a and other quinolones against MSSA and MRSA

Compound	X	R^1	R^6	R^7	Organism (number)						
					MSSA (48)			Quinolone-susceptible MRSA (46) [Quinolone-resistant MRSA (49)]			
					Range	MIC_{50}	MIC_{90}	Range	MIC_{50}	MIC_{90}	
DC-159a	C–OCH_3	(cyclopropyl-F)	F	(spiro aminopyrrolidine, H_2N)	0.015–1	0.03	0.06	0.015–0.06 [0.25–16]	0.03 [1]	0.06 [8]	
Ciprofloxacin	C–H	(cyclopropyl)	F	(piperazine, HN)	0.12–32	0.25	2	0.12–1 [8–>64]	0.5 [32]	0.5 [>64]	
Moxifloxacin	C–OCH_3	(cyclopropyl)	F	(bicyclic diamine)	0.015–2	0.06	0.12	0.015–0.06 [0.5–64]	0.03 [2]	0.06 [32]	

Table 5. Continued

Compound	X	R¹	R⁶	R⁷	Organism (number)						
					MSSA (48)			Quinolone-susceptible MRSA (46) [Quinolone-resistant MRSA (49)]			
					Range	MIC_{50}	MIC_{90}	Range	MIC_{50}	MIC_{90}	
Garenoxacin	C–OCHF₂	cyclopropyl	H	1-aminoisoindoline (methyl)	0.008–1	0.03	0.03	0.008–0.03 [0.25–32]	0.015 [1]	0.03 [32]	
Gemifloxacin	N	cyclopropyl	F	3-(aminomethyl)-4-(methoxyimino)pyrrolidine	0.008–1	0.03	0.12	0.008–0.06 [0.25–128]	0.03 [1]	0.03 [32]	
Levofloxacin	C–O–CH(CH₃)–	(fused)	F	4-methylpiperazine	0.06–8	0.25	0.5	0.12–0.5 [4–>128]	0.25 [8]	0.25 [>128]	

Table 6. Comparative MICs (μg/mL) of DC-159a and other quinolones against coagulase-negative staphylococci

Compound	X	R¹	R⁶	R⁷	Organism (number)						
					MSCNS (48)			MRCNS (46)			
					Range	MIC$_{50}$	MIC$_{90}$		Range	MIC$_{50}$	MIC$_{90}$
DC-159a	C–OCH$_3$	cyclopropyl-F	F	(aminomethyl spiro pyrrolidine)	0.015–1	0.03	0.5		0.03–8	0.25	0.5
Ciprofloxacin	C–H	cyclopropyl	F	piperazine	0.06–32	0.25	4		0.06–>64	4	32
Moxifloxacin	C–OCH$_3$	cyclopropyl	F	(diazabicyclo)	0.03–8	0.06	1		0.03–64	1	4

Table 6. Continued

Compound	X	R¹	R⁶	R⁷	Organism (number)						
					MSCNS (48)			MRCNS (46)			
					Range	MIC_{50}	MIC_{90}	Range	MIC_{50}	MIC_{90}	
Garenoxacin	C–OCHF₂	cyclopropyl	H	indoline	0.015–8	0.03	1	0.015–32	1	2	
Gemifloxacin	N	cyclopropyl	F	methoxyimino-pyrrolidinyl-methylamine	≤0.004–4	0.015	0.5	0.008–32	0.5	1	
Levofloxacin	C–O–CH(CH₃)–		F	methylpiperazinyl	0.06–32	0.25	4	0.12–>128	4	16	

Table 7. Comparative MICs (μg/mL) of DC-159a and other quinolones against MSSA and MRSA (nosocomial and community acquired isolates)

Compound	X	R¹	R⁷	Organism (number)					
				MSSA (30)			MRSA, nosocomial (30) [community acquired (30)]		
				Range	MIC_{50}	MIC_{90}	Range	MIC_{50}	MIC_{90}
DC-159a	C-OCH_3	-F (cyclopropyl)	(spiro pyrrolidine with H_2N)	0.015–0.06	0.03	0.03	0.5–4 [0.008–0.5]	1 [0.03]	2 [0.03]
Ciprofloxacin	C-H	(cyclopropyl)	(piperazine, HN)	0.06–0.5	0.25	0.5	>4 [≤0.25→4]	>4 [≤0.25]	>4 [0.5]
Moxifloxacin	C-OCH_3	(cyclopropyl)	(bicyclic HN, H)	≤0.03–0.12	≤0.03	0.06	1–>4 [≤0.03–1]	2 [≤0.03]	4 [0.06]

Table 7. Continued

Compound	X	R¹	R⁷	Organism (number)					
				MSSA (30)			MRSA, nosocomial (30) [community acquired (30)]		
				Range	MIC_{50}	MIC_{90}	Range	MIC_{50}	MIC_{90}
Gatifloxacin	C–OCH₃	cyclopropyl	HN–(3-methylpiperazinyl)	≤0.03–0.12	0.06	0.12	1→4 [≤0.03–2]	4 [0.06]	>4 [0.12]
Levofloxacin	C–O– (fused)	CH₃ (fused)	H₃CN–(piperazinyl)	0.06–0.25	0.12	0.25	4→8 [0.12–4]	>8 [0.12]	>8 [0.25]

Olamufloxacin (HSR-903)

Olamufloxacin (HSR-903) is a newly synthesized oral fluoroquinolone with a broad spectrum of antibacterial activity. It was first synthesized by Hokuriku Pharmaceutical Co. Ltd., Fukui, Japan [132, 133], and its chemical name is 5-amino-7-[(7S)-7-amino-5-azaspiro [2.4]heptan-5-yl]-1-cyclopropyl-6-fluoro-1,4-dihydro-8-methyl-4-oxoquinoline-3-carboxylic acid methanesulfonate.

Olamufloxacin (HSR-903)

The distinctive structural features of olamufloxacin are an amino moiety at the C-5 position, a 7-amino-5-azaspiro [2.4]heptan-5-yl ring at the C-7 position, and a methyl group at the C-8 position. In olamufloxacin, the 8-methyl group would contribute to the high hydrophobicity. In general, a 5-amino moiety, as found in sparfloxacin and DQ-113, enhances potency against Gram-positive bacteria. Few quinolones such as DQ-113, T-3912, have a methyl group at the C-8 position. The 7-amino-5-azaspiro [2.4]heptan-5-yl ring, as found in sitafloxacin, DK-507k, A-170568.1 and DC-159a is a cycloproyl-fused aminopyrrolidine ring. The aminopyrrolidine scaffold as found in tosufloxacin, clinafloxacin, and ABT-719 also enhances potency against Gram-positive bacteria [10]. In olamufloxacin, these substituents presumably contribute to the strong activity against Gram-positive bacteria.

The preliminary antibacterial data of olamufloxacin against MRSA were published by Takahashi et al. [134] and Watanabe et al. [135], respectively. In the comparative study of Takahashi et al., olamufloxacin inhibited 90% of MSSA and MRSA clinical isolates at 0.78 and 1.56 mg/mL, respectively, and its activity against MRSA was 16-fold higher than those of sparfloxacin and levofloxacin and 64-fold higher than that of ciprofloxacin (Table 8).

The in vitro activity of olamufloxacin against clinical isolates of respiratory pathogens was evaluated by Watanabe et al. [135] in comparison with ciprofloxacin, sparfloxacin and ofloxacin (Table 9).

Table 8. Comparative in vitro activities (MICs, µg/mL) of olamufloxacin (HSR-903) and other quinolones against clinical isolates of staphylococci

Compound	X	R^1	R^5	R^7	Organism (number)					
					MSSA (38)			MRSA (33)		
					Range	MIC_{50}	MIC_{90}	Range	MIC_{50}	MIC_{90}
Olamufloxacin (HSR-903)	C–CH$_3$	cyclopropyl	NH$_2$	(aminomethyl spiropyrrolidine)	0.006–6.25	0.025	0.78	0.006–3.13	0.39	1.56
Ciprofloxacin	C–H	cyclopropyl	H	(piperazinyl)	0.20–100	0.78	6.25	0.20–>100	12.5	100
Lomefloxacin	C–F	ethyl	H	(3-methylpiperazinyl)	0.39–>100	0.78	25	0.20–>100	25	>100

Table 8. Continued

Compound	X	R¹	R⁵	R⁷	Organism (number)					
					MSSA (38)			MRSA (33)		
					Range	MIC_{50}	MIC_{90}	Range	MIC_{50}	MIC_{90}
Sparfloxacin	C–F	cyclopropyl	NH_2	cis-3,5-dimethylpiperazinyl	0.025–25	0.10	12.5	0.012–50	3.13	25
Levofloxacin	C–O–CH(CH₃)–	(fused)	H	4-methylpiperazinyl	0.10–12.5	0.20	12.5	0.10–50	6.25	25

Table 9. Comparative in vitro activities (MICs, μg/mL) of olamufloxacin (HSR-903), ciprofloxacin, sparfloxacin and ofloxacin against clinical isolates of staphylococci

Compound	X	R^1	R^5	R^7	Organism (number) MSSA (20)			MRSA (20)		
					Range	MIC_{50}	MIC_{90}	Range	MIC_{50}	MIC_{90}
HSR-903	C–CH₃	cyclopropyl	NH₂	(aminomethyl-spiro-cyclopropyl pyrrolidine)	≤0.03–0.06	0.06	0.06	≤0.03–2	0.5	2
Ciprofloxacin	C–H	cyclopropyl	H	piperazinyl	0.12–1	0.25	0.5	2–>64	16	>64
Sparfloxacin	C–F	cyclopropyl	NH₂	3,5-dimethylpiperazinyl	0.06–0.12	0.06	0.12	0.06–16	8	16
Ofloxacin	C–O– (fused CH₃)	—	H	4-methylpiperazinyl	0.12–0.5	0.25	0.5	0.5–32	16	32

Olamufloxacin was 2 to 32 times more active than ofloxacin, ciprofloxacin, and sparfloxacin against Gram-positive bacteria, including MSSA and MRSA. One of the reasons why the antibacterial activity ratio of olamufloxacin to ciprofloxacin is greater for MRSA and ciprofloxacin-resistant *S. aureus* than that for MSSA is considered to be that olamufloxacin, with its high hydrophobicity, is little influenced by NorA. Quinolones with high hydrophobicity, such as sparfloxacin, are also reported to be active against a *norA*-mediated resistant strain [45, 136].

After oral administration to animals, olamufloxacin is well absorbed and is distributed into various tissues, including lung and kidney, but not the central nervous system [137]. It was demonstrated that olamufloxacin efficiently distributes in both epithelial lining fluid and alveolar macrophage, two potential and representative sites for pulmonary infections [138]. In the view of kinetic analysis, there existed the higher influx than efflux clearances across the alveolar barrier in favor of olamufloxacin penetration into epithelial lining fluid. Moreover, both rapid permeability across the alveolar macrophage cell membranes and avid binding to the membrane phospholipids were referred to account for the high accumulation of olamufloxacin within alveolar macrophage. In a toxicological study, olamufloxacin (30 mg/kg, intravenously) did not cause phototoxicity in guinea pigs [132].

DQ-113

DQ-113 (formerly D61-1113), 5-amino-7-[(3R,4S)-3-(1-aminocyclopropyl)-4-fluoropyrrolidin-1-yl]-6-fluoro-1-[(1R,2S)-2-fluorocyclopropyl]-1,4-dihydro-8-methyl-4-oxoquinoline-3-carboxylic acid is a highly functionalized fluoroquinolone.

DQ-113

This poly-fluorinated compound characterized by having three fluoro-substituents located on quinolone core, cyclopropyl ring and pyrrolidine moiety. Its N-1 substituent is similar to sitafloxacin. The 5-amino group (as found in sparfloxacin) and the 8-methyl substituent (as found in ABT-719, olamufloxacin and T-3912) presumably contribute to enhance potency against

Gram-positive pathogens. DQ-113 shares structural similarity with olamufloxacin based on possessing both 5-amino and 8-methyl groups.

Several in vitro and in vivo studies were reported with DQ-113. Tanaka *et al.* [139] published data on the antibacterial activity of DQ-113, in comparison with those of currently available quinolones, ciprofloxacin, gatifloxacin, gemifloxacin, levofloxacin, moxifloxacin, sitafloxacin, sparfloxacin, tosufloxacin, and garenoxacin (Tables 10 and 11). The MIC_{90}s of DQ-113 against MSSA and MSCNS were both 0.03 μg/mL. MIC_{90}s against ofloxacin-susceptible MRSA, ofloxacin-resistant (MIC of ofloxacin, ≥ 8 μg/mL) MRSA, and MRCNS were 0.008, 0.25, and 0.06 μg/mL, respectively. The antibacterial activity against MSSA was 2-fold higher than those of garenoxacin and gemifloxacin; 4-fold higher than those of sitafloxacin, tosufloxacin and moxifloxacin. Against ofloxacin-susceptible MRSA, DQ-113 activity was 4-fold higher than that of garenoxacin and at least 8-fold higher than those of the other reference compounds at MIC_{90} levels. Furthermore, DQ-113 showed the highest activity against ofloxacin-resistant MRSA among the compounds tested. Against MRCNS, the MIC_{90} of DQ-113 was at least 8-fold lower than those of the reference drugs.

Hong *et al.* compared DQ-113 respect to sitafloxacin, moxifloxacin, levofloxacin, and ciprofloxacin for potential to select mutational resistance in multi-resistant Gram-positive cocci. DQ-113 was the most effective agent for preventing the emergence of single-step mutants of MSSA and MRSA, with only one mutant selected from the 10 strains compared to mutants being selected from 4 strains by moxifloxacin, 5 strains by sitafloxacin, 7 strains by levofloxacin, and 8 strains by ciprofloxacin. In this study, all single-step mutants were susceptible to 0.03 μg/mL of DQ-113, 0.25 μg/mL of sitafloxacin, 0.5 μg/mL of moxifloxacin, 2 μg/mL of levofloxacin, and 8 μg/mL of ciprofloxacin, with frequencies of 10^{-6} and 10^{-10}, with no apparent differences between the quinolones [140].

Kaneko *et al.* [141] compared the effects of DQ-113 to those of vancomycin and teicoplanin in murine models of hematogenous pulmonary infections caused by MRSA and vancomycin-insensitive *S. aureus* (VISA). The MICs of DQ-113, vancomycin and teicoplanin for MRSA were 0.125, 1.0, and 0.5 μg/mL, respectively; and those for VISA were 0.25, 8.0, and 8.0 μg/mL, respectively. Kaneko *et al.* also concluded that DQ-113 effectively can reduce the number of bacteria in MRSA and VISA hematogenous infection models and significantly improves the rates of survival of immunocompromised mice infected with VISA compared with the rates achieved with vancomycin and teicoplanin. Thus, DQ-113 is potentially effective for the treatment of hematogenous pulmonary infections caused by MRSA and VISA strains.

Table 10. Comparative MICs (µg/mL) of DQ-113 and important quinolones against MSSA and MRSA

Compound	X	R^1	R^5	R^6	R^7	Organism (number)					
						MSSA (25)			MRSA, ofloxacin susceptible (24) [MRSA (25), ofloxacin resistant]		
						Range	MIC_{50}	MIC_{90}	Range	MIC_{50}	MIC_{90}
DQ-113	C–CH₃	(F-cyclopropyl)	NH₂	F	(aminopyrrolidinyl-F-cyclopropyl)	≤0.004–0.06	0.008	0.03	≤0.004–0.015 [0.03–2]	≤0.004 [0.06]	0.008 [0.25]
Sitafloxacin	C–Cl	(F-cyclopropyl)	H	F	(aminospiro)	0.008–0.12	0.03	0.12	0.008–0.06 [0.25–32]	0.03 [1]	0.06 [4]
Ciprofloxacin	C–H	(cyclopropyl)	H	F	(piperazinyl)	0.25–4	0.5	1	0.25–2 [16–>128]	0.5 [128]	1 [>128]
Gatifloxacin	C–OCH₃	(cyclopropyl)	H	F	(methylpiperazinyl)	0.06–0.5	0.12	0.25	0.06–0.25 [2–>128]	0.12 [8]	0.25 [16]

Table 10. Continued

Moxifloxacin	C–OCH₃	H	F	(octahydropyrrolopyridine)	0.06–0.25	0.06	0.12	0.03–0.25 [1–64]	0.12 [4]	0.12 [8]
Garenoxacin	C–OCHF₂	H	H	(methylindoline)	≤0.004–0.12	0.03	0.06	0.008–0.06 [0.25–64]	0.03 [2]	0.03 [8]
Gemifloxacin	N	H	F	(methoxyimino-pyrrolidine with CH₂NH₂)	0.03–0.25	0.06	0.06	0.015–0.06 [-]	0.06 [-]	0.06 [-]
Sparfloxacin	C–F	NH₂	F	(dimethylpiperazine)	0.06–0.25	0.12	0.25	0.03–0.12 [2→128]	0.06 [8]	0.12 [32]
Tosufloxacin	N	H	F	(aminopyrrolidine)	0.015–0.25	0.06	0.12	0.015–0.12 [1–64]	0.06 [32]	0.06 [32]
Levofloxacin	C–O–CH(CH₃)– (fused)	H	F	(methylpiperazine)	0.12–1	0.25	0.5	0.12–1 [4→128]	0.25 [16]	0.5 [64]

Table 11. Comparative MICs (μg/mL) of DQ-113 and important quinolones against coagulase-negative staphylococci

Compound	X	R^1	R^5	R^6	R^7	Organism (number)					
						MSCNS (27)			MRCNS (36)		
						Range	MIC_{50}	MIC_{90}	Range	MIC_{50}	MIC_{90}
DQ-113	C–CH$_3$	(cyclopropyl-F)	NH$_2$	F	(pyrrolidinyl, H$_2$N, cyclopropyl, F)	≤0.004–0.06	0.008	0.03	≤0.004–0.12	0.03	0.06
Sitafloxacin	C–Cl	(cyclopropyl-F)	H	F	(spiro pyrrolidine, H$_2$N)	0.008–0.25	0.03	0.25	0.008–0.5	0.12	0.5
Ciprofloxacin	C–H	(cyclopropyl)	H	F	(piperazinyl, HN)	0.06–64	0.25	8	0.12–128	8	64
Gatifloxacin	C–OCH$_3$	(cyclopropyl)	H	F	(methylpiperazinyl, HN, H$_3$C)	0.06–4	0.25	2	0.12–4	2	4

Table 11. Continued

Moxifloxacin	C–OCH$_3$	cyclopropyl	H	F	(octahydropyrrolo structure)	0.03–4	0.12	1	0.06–8	1	4
Garenoxacin	C–OCHF$_2$	cyclopropyl	H	H	(isoindoline structure)	0.015–4	0.06	1	0.03–4	1	4
Gemifloxacin	N	cyclopropyl	H	F	(pyrrolidine with H$_3$CO-N=, H$_2$N)	0.008–2	0.06	0.25	0.015–4	0.5	2
Sparfloxacin	C–F	cyclopropyl	NH$_2$	F	(H$_3$C, HN, H$_3$C piperazine)	0.03–8	0.12	4	0.06–16	4	8
Tosufloxacin	N	2,4-difluorophenyl	H	F	(H$_2$N-pyrrolidine)	0.015–16	0.06	4	0.03–16	4	16
Levofloxacin	C–O–CH(CH$_3$)	(fused)	H	F	(H$_3$CN piperazine)	0.12–8	0.25	4	0.12–32	4	16

Delafloxacin (ABT-492, RX-3341)

Delafloxacin (formerly A-319492, WQ-3034, ABT-492 and RX-3341) is a new fluoroquinolone that differs from other members of the class by two structural features: the 6-amino-3,5-difluoropyridine-2-yl moiety at the N-1 position and the 3-hydroxyazetidine-1-yl substituent at the C-7 position of the quinolone core. Delafloxacin is 1-(6-amino-3,5-difluoropyridine-2-yl)-8-chloro-6-fluoro-7-(3-hydroxyazetidine-1-yl)-4-oxo-1,4-dihydroquinoline-3-carboxylic acid, currently being developed by Rib-x Pharmaceutical Inc.

Delafloxacin
(ABT-492, RX-3341)

Preliminary data reported by Abbott Laboratories indicate that delafloxacin has a broad spectrum of activity with increased potency against multidrug-resistant bacteria [32]. Comparative study of the *in vitro* activity of delafloxacin published by Harnett et al. indicate that, this compound is significantly more active than other quinolones against staphylococci, but a higher MIC_{90} was observed for MRSA (0.5 μg/mL) compared with MSSA (0.008 μg/mL). Anti-staphylococcal activity of delafloxacin was superior for all species tested by Harnett et al., with good activity displayed against MRSA strains (MIC_{90} = 0.5 μg/mL) [142]. Delafloxacin has been shown to be at least 8-fold more potent than ciprofloxacin, gatifloxacin, moxifloxacin, gemifloxacin and levofloxacin (MIC_{90} of ≤ 0.5 μg/mL for delafloxacin versus > 4 μg/mL for all other quinolones) against ciprofloxacin resistant MRSA (Table 12).

In 2006, Rib-x Pharmaceutical Inc. licensed this antibacterial agent from a Japanese company, Wakunaga Pharmaceuticals, opening up a new antibiotic niche for Rib-x with invaluable clinical data. Based on activity against MRSA and other highly pathogenic microorganisms, delafloxacin has successfully undergone Phase II studies for oral treatment of respiratory infections, and

Phase I studies as an intravenous formulation. These studies have shown delafloxacin to be both clinically efficacious and safe. The oral administration of delafloxacin has been shown to have no evidence of either phototoxicity or QTc prolongation in Phase I studies and has demonstrated efficacy in two Phase II clinical trials (community acquired pneumonia and bronchitis when compared to levofloxacin) [143].

WCK 771

WCK 771, the L-arginine salt of *S*-(–)-nadifloxacin [*S*-(–)-9-fluoro-6,7-dihydro-8-(4-hydroxypiperidin-1-yl)-5-methyl-1-oxo-1*H*,5*H*-benzo[*i,j*] quinolizine-2-carboxylic acid L-arginine salt] (Figure 15) is a broad-spectrum tricyclic quinolone that is active against quinolone-resistant staphylococci and MRSA and is being studied in phase II clinical trials. *RS*-(±)-nadifloxacin was introduced only for topical use as a liniment against *Propionobacterium acnes* [144, 145]. Structurally, nadifloxacin has a lipophilic tricyclic benzoquinolizine nucleus, with a 4-hydroxypiperidinyl moiety at C-8 position. This 4-hydroxypiperidine moiety is a particular moiety without distal basic functionality, which is unusual for a side chain of quinolone. Most marketed quinolones bear side chains with basic nitrogen functionality, but this compound along with delafloxacin have a hydroxylcyclic amine without distal amine.

Treatment with *RS*-(±)-nadifloxacin has been reported to result in an unusually lower incidence of the emergence of resistant mutants. A Japanese study by Nishijima et al. against *S. aureus* isolates, including MRSA strains, collected during the period from 1994 to 2000 did not show significant increase in the MIC$_{90}$ of nadifloxacin [146, 147].

Similar to the case of levofloxacin [(*S*)-isomer] and ofloxacin (racemic), the levorotatory (*S*)-isomer of nadifloxacin is 64–256 times more potent than its *R*-(+)-isomer and approximately twice as active as the *RS*-(±)-nadifloxacin against Gram-positive and -negative bacteria [144, 148]. Because the *S*-(–)-isomer is primarily responsible for antibacterial activity, the potency of WCK 771 [arginine salt of the *S*-(–) isomer] is two to four times higher than that of racemic nadifloxacin [149].

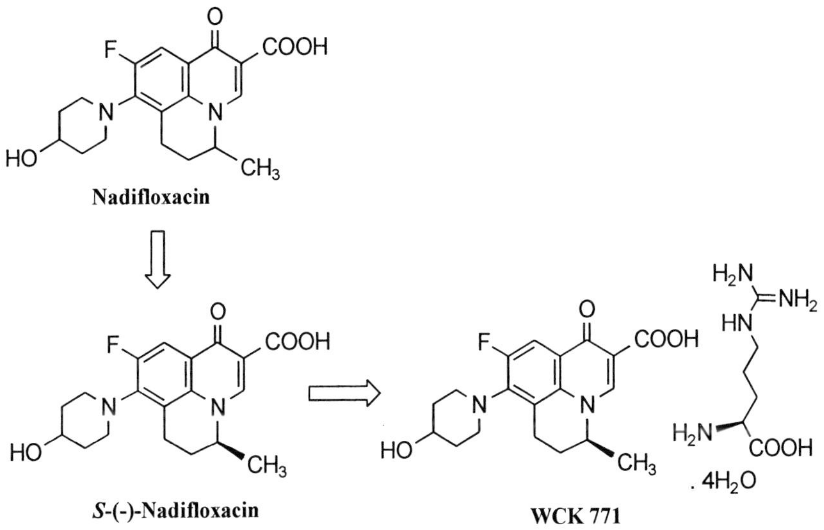

Figure 15. Development of WCK 771.

The unusual physicochemical properties of S-(−)-nadifloxacin, such as high partition coefficients, low aqueous solubility and having a single pKa value distinct it from representative quinolones. However, it is envisaged that the less aqueous solubility may be compensated by its high potency coupled with the lipophilicity-driven permeability, which makes S-(−)-nadifloaxcin optimizable for an oral administration. Alternatively, it is also envisaged that a water-soluble salt form of S-(−)-nadifloaxcin can be considered for injection preparation. Various prodrugs and salts of S-(−)-nadifloaxcin were studied by de Souza *et al.* and the L-arginine salt of S-(−)-nadifloxacin (WCK 771) was selected as an entity for further development as a potential drug candidate [150].

The comparative study of WCK 771 with other quinolones demonstrates that this new agent is a highly potent anti-staphylococcal quinolone with improved potency against even quinolone-resistant strains of *S. aureus* and coagulase-negative staphylococci. In Patel *et al.* study, the potency of WCK 771 against 176 strains of MRSA (MIC_{90} = 1.0 µg/mL) was considerably higher than that of levofloxacin (MIC_{90} = 16 µg/mL) or ciprofloxacin (MIC_{90} = 64 µg/mL). It was two- to fourfold more potent than moxifloxacin and trovafloxacin (Table 13). Additionally, WCK 771 has potency comparable to that of clinafloxacin [151].

Table 12. Comparative MICs (μg/mL) of delafloxacin and other quinolones against MSSA and MRSA

Compound	X	R¹	R⁷	Organism (number)						
				MSSA (50)			MRSA (25)			
				Range	MIC_{50}	MIC_{90}	Range	MIC_{50}	MIC_{90}	
Delafloxacin	C–Cl	2,4-difluoro-6-amino-pyridinyl	3-hydroxyazetidinyl	0.001–0.12	0.004	0.008	0.008–1	0.5	0.5	
Ciprofloxacin	C–H	cyclopropyl	piperazinyl	0.12–128	0.5	1	0.5–>128	64	128	
Gatifloxacin	C–OCH₃	cyclopropyl	3-methylpiperazinyl	0.06–8	0.12	0.25	0.25–32	4	8	

Table 12. Continued

Compound	X	R^1	R^7	Organism (number)					
				MSSA (50)			MRSA (25)		
				Range	MIC_{50}	MIC_{90}	Range	MIC_{50}	MIC_{90}
Moxifloxacin	C–OCH$_3$	cyclopropyl	(octahydropyrrolo-pyridine)	0.03–2	0.06	0.12	0.12–8	2	4
Gemifloxacin	N	cyclopropyl	(methoxyimino-pyrrolidinyl-methylamine)	0.015–8	0.03	0.06	0.03–16	8	8
Levofloxacin	C–O–CH(CH$_3$)		(methylpiperazinyl)	0.12–16	0.25	0.5	0.5–128	16	16

Table 13. Comparative MICs (μg/mL) of WCK 771 and representative quinolones against clinical isolates of staphylococci

Compound	X	R¹	R⁷	Organism (number)					
				MSSA (244)			MRSA (176)		
				Range	MIC_{50}	MIC_{90}	Range	MIC_{50}	MIC_{90}
WCK 771	C	(S)-sec-butyl with CH₃	4-hydroxypiperidinyl	0.015–0.25	0.03	0.03	0.015–4.0	0.5	1.0
Levofloxacin	C–O	(S)-CH(CH₃)-	3-methylaminopiperazinyl	0.12–1.0	0.25	0.5	0.125–64	4.0	16.0
Ciprofloxacin	C–H	cyclopropyl	piperazinyl	0.12–2.0	0.5	1.0	0.25–128	4.0	64
Clinafloxacin	C–Cl	cyclopropyl	3-aminopyrrolidinyl	0.015–0.25	0.03	0.06	0.015–4.0	0.25	1.0

Table 13. Continued

Compound	X	R¹	R⁷	Organism (number)						
				MSSA (244)			MRSA (176)			
				Range	MIC$_{50}$	MIC$_{90}$	Range	MIC$_{50}$	MIC$_{90}$	
Moxifloxacin	C–OCH$_3$	(cyclopropyl)	(octahydropyrrolo-pyridinyl)	0.03–0.5	0.03	0.12	0.03–8.0	1.0	4.0	
Trovafloxacin	N	(2,4-difluorophenyl)	(azabicyclo amine)	0.015–0.5	0.06	0.12	0.015–16	1.0	4.0	

Also, Patel et al. evaluated the potency of WCK 771 in animal models of staphylococcal infection. WCK 771 was effective by both the oral and the subcutaneous routes in mice infected intraperitoneally with quinolone-susceptible MSSA strains. For infections caused by quinolone-resistant MRSA strains, the activity of WCK 771 administered subcutaneously was superior to those of trovafloxacin and sparfloxacin, with an ED_{50} range of 27.8 to 46.8 mg/kg. The activity of WCK 771 was superior to those of moxifloxacin, vancomycin, and linezolid in a mouse cellulitis model of infection caused by one strain of MSSA and two strains of MRSA, with effective doses of 2.5 and 5 mg/kg for the MSSA strain and 10-fold higher effective doses for MRSA strains. WCK 771, like vancomycin and linezolid, eradicated MRSA from mouse liver, spleen, kidney, and lung when it was administered subcutaneously at a dose of 50 mg/kg for four doses [151].

In assessing the impact of the NorA efflux pump on WCK 771 activity by Patel et al., it was observed that the activity of the drug against a NorA-hyperexpressing *S. aureus* strain was not affected by the presence of reserpine and that there was little difference between the MICs for $NorA^-$ strains and those for $NorA^+$ [151].

The strong potency of WCK 771 against MSSA and MRSA strains is possibly due to its unique mechanism of action and its resistance to efflux-mediated resistance, resulting in the combination of high-affinity targeting of staphylococcal DNA gyrase, coupled with efflux resistance, creating high intracellular drug concentrations.

To estimate the range of target mutations that may affect the activity of WCK 771 and to determine its primary and secondary targeting properties, Bhagwat et al. characterized sequentially selected staphylococcal mutants and determined their quinolone susceptibilities. WCK 771 was compared with other quinolones possessing improved anti-staphylococcal activities and diverse targeting properties, such as moxifloxacin, trovafloxacin, and garenoxacin [152]. There was a 2-fold increase in the MICs of WCK 771 for single *gyrA* mutants, indicating that DNA gyrase is its primary target. All first- and second-step mutants selected by WCK 771 revealed *gyrA* and *grlA* mutations, respectively. The MICs of WCK 771 and clinafloxacin were found to be superior to those of other quinolones against strains with double and triple mutations. This study demonstrated that *gyrA* is the primary target of WCK 771 and its target preference is remarkably different from those of quinolones with dual targets (garenoxacin and moxifloxacin) and topoisomerase IV-specific quinolones (trovafloxacin). WCK 771 displays an activity profile comparable to that of clinafloxacin, a dual-acting quinolone with a high affinity to DNA gyrase.

Based on the pharmacological profile of WCK 771, it could be considered as a drug candidate for use against nosocomial multidrug-resistant Gram-positive infections, in particular MRSA and VRSA infections.

6-FLUORONAPHTHYRIDONES

Gemifloxacin

Gemifloxacin (SB-265805, LB20304a) possessing excellent activity against streptococci and staphylococci, is a fluoronaphthyridone derivative with a 4-(aminomethyl)-3-(methoxyimino)pyrrolidine substituent at the C-7 position. Its N-1 side chain includes a cyclopropyl group characteristic of safer fluoroquinolones in contrast to the difluorophenyl substituent of other naphthyridones, tosufloxacin and trovafloxacin. Gemifloxacin carries a methoxyimino- group with (Z)-configuration on C-7 pyrrolidine ring. This compound is a racemic (R, S) mixture.

Gemifloxacin

Increased bulkiness of C-7 position appears to confer protection from efflux exporter proteins and diminishes the likelihood of resistance in wild-type strains [69]. Replacement of the carbon at the C-8 position with nitrogen enhances antibacterial potency, apparently due to the increased activity against both DNA gyrase and topoisomerase IV so that ≥ 2 mutations in the QRDRs are usually required for drug resistance to develop.

Preliminary evaluation of gemifloxacin against MSSA indicated that gemifloxacin was equiactive to trovafloxacin ($MIC_{90}s = 0.125$ µg/mL) and more active than comparator agents ($MIC_{90} = 1$ µg/mL for ofloxacin; $MIC_{90} = 2$ µg/mL for ciprofloxacin). The activity of gemifloxacin against MRSA was equal to trovafloxacin ($MIC_{90}s = 4$ µg/mL). However, gemifloxacin was more potent against MRSA than other quinolones ($MIC_{90}s$ = 32 μg/mL for grepafloxacin, ofloxacin, and ciprofloxacin) [153].

In another study, gemifloxacin demonstrated better antibacterial activity against 39 strains of *S. aureus* isolated from patients with sinusitis than levofloxacin, gatifloxacin, and moxifloxacin, with an MIC_{90} <0.03 μg/mL compared with 0.25, 0.125, and 0.125 μg/mL of comparator drugs, respectively [154]. Against 346 contemporary clinical isolates of methicillin and ciprofloxacin-sensitive *S. aureus* in the US and Canada, gemifloxacin demonstrated 0.03 μg/mL compared with 1–16 μg/mL for amoxicillin/clavulanate and azithromycin. Gemifloxacin, however, failed to show efficient antibacterial activity against 41 methicillin-resistant ATCC strains of *S. aureus* and 160 methicillin- and ciprofloxacin-resistant clinical isolates from 11 North American medical centers; MIC_{90} values against the 2 groups of MRSA were 2 and 16 μg/mL, respectively [155, 156].

DW286

DW286, 7-[3-(aminomethyl)-4-(methoxyimino)-3-methylpyrrolidin-1-yl]-1-cyclopropyl-6-fluoro-4-oxo-1,4-dihydro[1,8]naphthyridine-3-carboxylic acid hydrochloric acid salt, is a fluoronaphthyridone developed by Dong Wha (South Korea). DW286 is methyl analog of gemifloxacin. In fact, the pyrrolidine ring of gemifloxacin has been substituted by a methyl group at position 3, resulted in DW286. Structurally, gemifloxacin and DW286 are distinguished from other pyrrolidin-1-ylquinolones by attaching a methoxyimino group on pyrrolidine ring.

The in vitro and in vivo activities of DW286 have been compared with those of ciprofloxacin, gemifloxacin, sparfloxacin, and trovafloxacin (Table 14). DW286 has potent antibacterial activity against staphylococci, including MRSA and quinolone-resistant *S. aureus* (QRSA).

Table 14. Efficacy of DW286 in comparison with ciprofloxacin, gemifloxacin, sparfloxacin and trovafloxacin on systemic infections in mice

Compound	X	R^1	R^5	R^7	S. aureus Smith [1.2×10^7]		MRSA MB4-19 [3.1×10^8]		QRSA MB4-20 [5.0×10^8]	
					MIC (µg/ml)	PD_{50} (mg/kg)	MIC (µg/ml)	PD_{50} (mg/kg)	MIC (µg/ml)	PD_{50} (mg/kg)
DW286	N	cyclopropyl	H	H₃CO-N=, CH₃, CH₂NH₂	0.004	0.12	0.004	0.12	0.25	5.83
Gemifloxacin	N	cyclopropyl	H	H₃CO-N=, CH₂NH₂	0.031	0.77	0.016	0.70	2	>40

Table 14. Continued

Compound	X	R^1	R^5	R^7	Microorganism [challenge dose (CFU/ml)]					
					S. aureus Smith [1.2×10^7]		MRSA MB4-19 [3.1×10^8]		QRSA MB4-20 [5.0×10^8]	
					MIC (µg/ml)	PD_{50} (mg/kg)	MIC (µg/ml)	PD_{50} (mg/kg)	MIC (µg/ml)	PD_{50} (mg/kg)
Ciprofloxacin	C–H	cyclopropyl	H	piperazinyl	0.5	3.43	0.25	3.99	32	>40
Sparfloxacin	C–F	cyclopropyl	NH_2	3,5-dimethylpiperazinyl	0.063	0.54	0.031	0.22	32	>40
Trovafloxacin	N	2,4-difluorophenyl	H	azabicyclic amine	0.016	0.32	0.016	0.17	8	>40

The in vitro activity of DW286 was stronger than that of mentioned quinolones. In an experimental mouse model of systemic infection caused by three *S. aureus* strains, including MRSA and QRSA, DW286 demonstrated the most potent activity, as found in vitro. DW286 was >8-fold more active against QRSA than the other quinolones [157]. It also exhibited better in vitro activity than several quinolones against mutant strains that developed quinolone resistance in vitro and was a poor substrate for efflux pumps [158].

Zabofloxacin (DW-224a)

Zabofloxacin (formerly DW-224a) is a new fluoroquinolone with the following chemical formula: 1-cyclopropyl-6-fluoro-7-[8-(methoxyimino)-2,6-diazospiro[3,4]oct-6-yl]-4-oxo-1,4-dihydro[1,8]naphthyridine-3-carboxylic acid hydrochloride [159, 160]. Zabofloxacin synthesized by the research laboratory of Dong Wha Pharmaceutical Industry, Ltd. (Anyang, Republic of Korea).

Zabofloxacin
(DW-224a)

Different studies demonstrated that zabofloxacin has a potent in vitro and in vivo antibacterial activity against both Gram-positive and Gram-negative bacteria. The in vitro efficacy of zabofloxacin in comparison with those of ciprofloxacin, moxifloxacin, and gemifloxacin has been demonstrated by Park et al. (Table 15) [161]. Zabofloxacin was slightly superior to gemifloxacin and much more active than moxifloxacin and ciprofloxacin against clinical isolates of Gram-positive bacteria. Against MSSA and MRSA, the antibacterial activity of zabofloxacin was comparable to that of gemifloxacin but 2- to 16-fold more potent than those of moxifloxacin and ciprofloxacin (Table 15). The MIC_{90}s of zabofloxacin against MSSA and MRSA were 0.03 μg/mL and 4 μg/mL, respectively. Against coagulase-negative staphylococci (both MSCNS and MRCNS), the activity of zabofloxacin was also 2- to 16-fold more potent than those of gemifloxacin, ciprofloxacin and moxifloxacin (Table 16).

Table 15. Comparative MICs (μg/mL) of zabofloxacin (DW-224a), gemifloxacin, ciprofloxacin and moxifloxacin against MSSA and MRSA

Compound	X	R⁷	Organism (number)						
			MSSA (62)			MRSA (120)			
			Range	MIC_{50}	MIC_{90}	Range	MIC_{50}	MIC_{90}	
Zabofloxacin (DW-224a)	N	H₃CO-N=... (spiro pyrrolidine-azetidine)	0.008–0.06	0.015	0.03	0.008–32	2	4	
Gemifloxacin	N	H₃CO-N=... CH₂NH₂ pyrrolidine	0.008–0.06	0.015	0.03	0.008–64	2	4	
Ciprofloxacin	C-H	piperazine	0.06–0.5	0.25	0.5	0.125–>64	32	>64	
Moxifloxacin	C-OCH₃	octahydropyrrolo[3,4-b]pyridine	0.015–0.25	0.06	0.06	0.03–>64	4	16	

Table 16. Comparative MICs (μg/mL) of zabofloxacin (DW-224a), gemifloxacin, ciprofloxacin and moxifloxacin against coagulase-negative staphylococci

Compound	X	R⁷	Organism (number)						
			MSCNS (14)			MRCNS (38)			
			Range	MIC$_{50}$	MIC$_{90}$	Range	MIC$_{50}$	MIC$_{90}$
Zabofloxacin (DW-224a)	N	H₃CO-N=... (spiro pyrrolidine-azetidine with HN)	0.008–0.25	0.015	0.125	0.008–4	0.25	2
Gemifloxacin	N	H₃CO-N=... pyrrolidine with H₂N-CH₂	0.008–0.25	0.015	0.25	0.008–8	0.25	4
Ciprofloxacin	C-H	HN-piperazine-N	0.125–2	0.125	2	0.06–64	8	32
Moxifloxacin	C-OCH₃	octahydropyrrolo[3,4-b]pyridinyl	0.03–2	0.125	0.25	0.06–16	2	8

In the systemic infection model caused by *S. aureus* giorgio, the ED_{50}s of zabofloxacin, ciprofloxacin, moxifloxacin and gemifloxacin were 1.11, 16.52, 1.78, and 1.37 mg/kg of body weight, respectively. Especially against quinolone-resistant *S. aureus* P197 (an MRSA), zabofloxacin was more effective than ciprofloxacin, moxifloxacin, and gemifloxacin [161].

The safety of zabofloxacin after oral administration has been studied in mice, rats and dogs [162]. Zabofloxacin appears to have exerted no adverse effects on the central nervous, cardiovascular and respiratory systems with the exception of the effect on the QT interval prolongation [163].

6-DESFLUOROQUINOLONES

Garenoxacin (T-3811, BMS-284756)

Garenoxacin (T-3811, BMS-284756) is a new 6-desfluoroquinolone that was developed by Bristol-Myers Squibb (New York) under license from Toyama Chemical Co. (Toyama, Japan). Bristol-Myers Squibb returned all rights for the drug to Toyama in 2003 and an agreement was signed in 2004 [164]. Its structural feature is the absence of fluorine at C-6 but has fluorine incorporated through a C-8 difluoromethoxy group. In addition, garenoxacin has a certain aryl moiety (isoindolin-5-yl) instead of cyclic amine at C-7, which connected to the quinolone core via carbon-carbon bond [165, 166].

Garenoxacin

Garenoxacin has been evaluated for treatment of respiratory, urinary tract, and skin and soft tissue infections caused by susceptible and resistant pathogens [164]. This non-fluorinated quinolone is reported to be active against MRSA strains that are resistant to other fluoroquinolones and selects fluoroquinolone-resistant mutants at a lower frequency than older quinolones [157, 167]. Antibacterial studies have indicated that MSSA and MRSA were susceptible to garenoxacin at MIC breakpoint of 0.5 μg/mL and 1 μg/mL, respectively. Based on studies with reserpine as an efflux pump inhibitor, the presence of reserpine did not result in a decrease in garenoxacin MICs for any of the isolates tested. Thus, garenoxacin is not a substrate for efflux pump activity via NorA [38, 168]. However, a recent study of *S. aureus* clinical isolates in the Asia-Pacific region indicated that 1% and 9% of MSSA and MRSA isolates resistant to ciprofloxacin were also resistant to 4 μg/mL garenoxacin, respectively [165].

DX-619

DX-619 is a new 6-desfluoroquinolone which was developed as a drug used for the treatment of infections caused by multidrug-resistant Gram-positive organisms. Its chemical name is 7-[(3R)-3-(1-aminocyclopropyl)-pyrrolidin-1-yl]-1-[(1R,2S)-2-fluorocyclopropyl]-8-methoxy-1,4-dihydro-4-oxoquinoline-3-carboxylic acid.

DX-619

DX-619 shares structural similarity with DK-507k, DC-159a based on possessing both 2-fluorocyclopropyl at N-1 and methoxy group at C-8. A certain pyrrolidine ring bearing aminocyclopropyl pendent at C-7 position, differentiates DX-619 from earlier quinolones and 6-desfluoroquinolones. Indeed, the aminomethyl residue which is found in gemifloxacin and DW286, fused with cylopropane in DX-619 resulted in aminocyclopropyl pendent. Development of 6-desfluoroquinolones such as DX-619 demonstrated that the impact of 6-fluoro- substituent is diminished when the molecule contains spatial arrangements with other appropriate substituents.

Watanabe *et al.* [169] tested susceptibilities of MRSA isolates from both hospitals and the community to DX-619 in comparison with sitafloxacin, levofloxacin, and other anti-MRSA antibiotics. They used different categories of strains: healthcare-associated MRSA (HA-MRSA) strains, community-associated MRSA (CA-MRSA)/non-multiresistant oxacillin-resistant *S. aureus* (NORSA) strains; healthcare-associated methicillin-resistant coagulase-negative staphylococci (HA-MRCNS) strains and community-associated methicillin-resistant coagulase-negative staphylococci (CA-MRCNS) strains. DX-619 showed the lowest MIC_{90} values for all categories of strains tested, irrespective of the degree of glycopeptide resistance. The six strains with MIC values of >128 μg/mL of levofloxacin commonly carried two mutations in *gyrA* and two mutations in *grlA*. DX-619 showed potent activity against strains with MIC values of 2 μg/mL.

In a Japanese study, Yamamoto et al. [170] surveyed and characterized highly virulent, community-acquired MRSA strains with Panton-Valentine leucocidin (PVL) genes (PVL^+ strains of MRSA) isolated from patients in hospitals. They also examined the in vitro susceptibilities of PVL^+ MRSA strains to 31 antibacterial agents, including DX-619 and sitafloxacin. DX-619 showed the greatest activity against PVL^+ MRSA among the antibacterial agents tested (MIC_{50} and MIC_{90} = 0.008 μg/mL). This activity was 8-fold greater than that of sitafloxacin (MIC_{90} = 0.063 μg/mL), 32-fold greater than that of levofloxacin (MIC_{90} = 0.25 μg/mL), and 125-fold greater than that of vancomycin (MIC_{90} = 1 μg/mL).

Bogdanovich et al. [171] examined the activity of DX-619 against 259 methicillin- and quinolone-susceptible and –resistant staphylococci compared to those of ciprofloxacin, gatifloxacin, moxifloxacin, sitafloxacin and levofloxacin (Tables 17 and 18). DX-619 had the lowest MICs against 131 *S. aureus* strains (MIC_{90} = 0.5 μg/mL) and 128 coagulase-negative staphylococci (MIC_{90} = 0.125 μg/mL). Among strains tested, 76 *S. aureus* strains and 51 coagulase-negative staphylococci were resistant to ciprofloxacin. DX-619 had the lowest MIC_{90} values against 127 quinolone-resistant staphylococci (0.5 μg/mL), followed by sitafloxacin (4 μg/mL), moxifloxacin (8 μg/mL), gatifloxacin (16 μg/mL), levofloxacin (32 μg/mL), and ciprofloxacin (>32 μg/mL). DX-619 and sitafloxacin were also more active than other tested drugs against selected mutants and had the lowest mutation frequencies in single-step resistance selection.

According to Strahilevitz et al. studies against wild-type *S. aureus* [172], DX-619 was 16- to 32-fold, twofold, and four- to eightfold more potent than ciprofloxacin, gemifloxacin, and garenoxacin, respectively. DX-619 manifested equal fourfold increases in MIC against a common *parC* mutant and a common *gyrA* mutant and selected for mutants at up to two- to fourfold its MIC, consistent with dual-targeting properties. Studies with purified topoisomerase IV and gyrase from *S. aureus* also showed that DX-619 had similar activity against topoisomerase IV and gyrase.

Table 17. Comparative MICs (μg/mL) of DX-619 and important quinolones against MSSA and MRSA

[Structure: quinolone core with R¹ on N, R⁶ and R⁷ on ring, X position, COOH at 3-position, 4-oxo]

Compound	X	R¹	R⁶	R⁷	Organism (number)						
					MSSA (69)			MRSA (62)		
					Range	MIC$_{50}$	MIC$_{90}$	Range	MIC$_{50}$	MIC$_{90}$
DX-619	C–OCH$_3$	fluorocyclopropyl	H	1-amino-cyclopropyl-pyrrolidinyl	≤0.002–0.5	0.008	0.125	0.008–2.0	0.125	1.0
Ciprofloxacin	C–H	cyclopropyl	F	piperazinyl	≤0.06–>32.0	0.5	>32.0	0.5–>32.0	>32.0	>32.0
Gatifloxacin	C–OCH$_3$	cyclopropyl	F	3-methylpiperazinyl	0.03–16.0	0.125	8.0	0.125–>32.0	8.0	16.0
Moxifloxacin	C–OCH$_3$	cyclopropyl	F	octahydropyrrolo-pyridinyl	≤0.016–8.0	0.06	4.0	0.06–>32.0	2.0	16.0

Table 17. Continued

Compound	X	R¹	R⁶	R⁷	Organism (number)						
					MSSA (69)			MRSA (62)			
					Range	MIC_{50}	MIC_{90}	Range	MIC_{50}	MIC_{90}	
Sitafloxacin	C–Cl	fluorocyclopropyl	F	7-amino-5-azaspiro[2.4]heptyl	≤0.008–8.0	0.06	1.0	0.03–32.0	1.0	8.0	
Levofloxacin	C–O (methyl-bridged)		F	3-methylpiperazinyl	0.06–>32.0	0.25	16.0	0.25–>32.0	16.0	>32.0	

Table 18. Comparative MICs (μg/mL) of DX-619 and important quinolones against coagulase-negative staphylococci

Compound	X	R^1	R^6	R^7	Coagulase-negative staphylococci (number)						
					MSSA (67)			MRSA (61)			
					Range	MIC_{50}	MIC_{90}	Range	MIC_{50}	MIC_{90}	
DX-619	C–OCH$_3$	cyclopropyl-F	H	aminomethyl-pyrrolidinyl	0.004–0.25	0.016	0.06	0.004–0.25	0.06	0.125	
Ciprofloxacin	C–H	cyclopropyl	F	piperazinyl	≤0.06–>32.0	0.25	8.0	0.25–>32.0	16.0	>32.0	
Gatifloxacin	C–OCH$_3$	cyclopropyl	F	3-methylpiperazinyl	0.06–4.0	0.125	2.0	0.06–32.0	2.0	4.0	

Table 18. Continued

Compound	X	R¹	R⁶	R⁷	Coagulase-negative staphylococci (number)						
					MSSA (67)			MRSA (61)			
					Range	MIC_{50}	MIC_{90}	Range	MIC_{50}	MIC_{90}	
Moxifloxacin	C–OCH₃	(cyclopropyl)	F	(bicyclic diamine)	0.03–4.0	0.125	1.0	0.03–8.0	1.0	4.0	
Sitafloxacin	C–Cl	(fluorocyclopropyl)	F	(spiro aminopyrrolidine)	0.016–0.25	0.03	0.25	0.016–2.0	0.25	0.5	
Levofloxacin	C–O–CH(CH₃)–		F	(methylpiperazine)	0.125–16.0	0.25	8.0	0.125–>32	8.0	16.0	

Susceptibility studies with DX-619 and an array of efflux pump substrates with and without reserpine, an inhibitor of efflux pumps, suggested that resistance in DX-619-selected mutants is affected by mechanisms other than mutations in topoisomerases or known reserpine-inhibitable pumps in *S. aureus* and thus are likely novel [172].

In general, DX-619 is a potent agent against both MSSA and MRSA, and coagulase-negative staphylococci, suggesting that it would be a promising candidate for the treatment of MRSA infection if sufficient in vivo concentrations were safely attained. Preliminary studies by Daiichi Sankyo Co. Ltd. have suggested that DX-619 has a promising safety profile [169].

Nemonoxacin (TG-873870)

Nemonoxacin (TG-873870) is a 6-desfluorinated quinolone, developed by TaiGen Biotechnology Co. (Taipei, Taiwan). Its chemical name is 7-[(3*S*,5*S*)-3-amino-5-methylpiperidin-1-yl]-1-cyclopropyl-8-methoxy-4-oxo-1,4-dihydroquinoline-3-carboxylic acid [173].

Nemonoxacin (TG-873870)

Beside the absence of fluorine atom at C-6, the distinctive structural feature of nemonoxacin is a (3*S*,5*S*)-3-amino-5-methylpiperidin-1-yl moiety at C-7. The most frequent substitution of quinolones at C-7 is piperazine or pyrrolidine moiety. Thus quinolones with cyclic amine at C-7 can categorize as piperazin-1-yl- and pyrrolidin-1-yl- quinolones. The third category is piperidin-1-ylquinolones. Nemonoxacin along with WCK 771, WCK 1152 and WCK 1153 are piperidin-1-ylquinolone. Also, the N-1-cyclopropyl and C-8-methoxy substituents are preserved in this molecule similar to gatifloxacin, moxifloxacin, WCK 1152 and WCK 1153.

Nemonoxacin has good activity against MRSA, particularly toward community-acquired MRSA. Also, This agent has an excellent activity toward vancomycin-intermediate or resistant *S. aureus*. In a comparative in vitro study with other quinolones in clinical use, nemonoxacin is less prone to the development of resistance and is active against quinolone-resistant clinical strains [174].

Ozenoxacin (T-3912)

Ozenoxacin (T-3912) is a new non-fluorinated quinolone (similar to garenoxacin), developed by Toyama Chemical Co. Its chemical name is 1-cyclopropyl-8-methyl-7-[5-methyl-6-(methylamino)-3-pyridinyl]-4-oxo-1,4-dihydro-3-quinolinecarboxylic acid [175].

Ozenoxacin (T-3912)

Its structural modifications respect to typical quinolones is the absence of 6-fluoro, the presence of 8-methyl and the attachment of certain aminoaryl moiety instead of aminocyclic amine to the C-7 position. Among the quinolones, there are only a limited number of molecules in which substituent attached to C-7 through carbon-carbon linkage. In ozenoxacin similar to garenoxacin, the C-7 aryl ring is connected to the quinolone core via carbon-carbon bond.

The antibacterial activity of ozenoxacin has been compared with that of nadifloxacin, ofloxacin, levofloxacin, clindamycin, erythromycin and gentamicin (Table 19) [176]. The *in vitro* activity of ozenoxacin against MSSA, ofloxacin-resistant MRSA was four-fold to 16000-fold greater than that of other agents at the MIC_{90} for the clinical isolates. Yamakawa et al. compared the inhibitory activity of ozenoxacin with other quinolones for DNA gyrase and topoisomerase IV of *S. aureus* SA113. Ozenoxacin showed the greatest inhibitory activity for both enzymes among the quinolones tested. This study showed that the activity of ozenoxacin was not influenced by *grlA* mutation in *S. aureus*, and the degree of MIC increase of ozenoxacin for *grlA-gyrA* double and triple mutants was lowest among the quinolones tested (nadifloxacin, levofloxacin and ofloxacin). Yamakawa et al. concluded that ozenoxacin is potentially a useful quinolone for the treatment of skin and soft-tissue infections and that its potent bactericidal activity might be able to shorten the treatment period.

Table 19. Antibacterial activity (MIC, μg/mL) of ozenoxacin (T-3912) and other antibacterial agents against clinical isolates of staphylococci

Compound	Organism (number)					
	MSSA (25)			MRSA, ofloxacin-resistant (23)		
	Range	MIC_{50}	MIC_{90}	Range	MIC_{50}	MIC_{90}
Ozenoxacin (T-3912)	0.00313–0.00625	0.00625	0.00625	0.025–0.2	0.2	0.2
Nadifloxacin	0.0125–0.05	0.025	0.05	0.78–1.56	1.56	1.56
Ofloxacin	0.2–1.56	0.39	0.78	12.5–100	50	50
Levofloxacin	0.1–0.39	0.2	0.39	3.13–50	25	25
Clindamycin	0.1–>100	0.1	0.2	>100	>100	>100
Erythromycin	0.2–>100	0.2	>100	>100	>100	>100
Gentamicin	0.05–100	0.39	100	0.2–>100	25	100

QUINAZOLINEDIONES

PD 0305970 and PD 0326448

Quinazoline-2,4-diones such as PD 0305970 and PD 0326448 are new bacterial type II topoisomerase inhibitors structurally related to the quinolones. The chemical name of PD 0305970 is 3-amino-7-{(R)-3-[(S)-1-amino-ethyl]-pyrrolidin-1-yl}-1-cyclopropyl-6-fluoro-8-methyl-1H-quinazoline-2,4-dione. PD 0326448 is the 3-desamino analog of PD 0305970.

PD 0305970

PD 0326448

Historically, quinazolinediones as antibacterial appeared in the literature as early as 1961, but it was not until 2004 that researchers at Pfizer reported similar antibacterial agents [177]. PD 0305970 and PD 0326448 were developed to introduce an orally active quinazolinedione, displaying antibacterial activity against fastidious Gram-negative and Gram-positive organisms, including multidrug- and fluoroquinolone-resistant organisms. In Huband et al. study, the respective MIC_{90} values against staphylococci for PD 0305970 were from 0.125 to 0.5 µg/mL. The MIC_{90}s of PD 0326448 were generally twofold higher than that of PD 0305970 (Table 20) [178, 179].

Table 20. Comparative antibacterial activities (MICs, µg/mL) of PD 0305970, PD 0326448, garenoxacin, gatifloxacin and levofloxacin against clinical isolates of staphylococci

Compound	*Staphylococcus aureus* (number)								
	Oxacillin susceptible (20)			Oxacillin resistant (47)			Oxacillin resistant, levofloxacin resistant (35)		
	Range	MIC_{50}	MIC_{90}	Range	MIC_{50}	MIC_{90}	Range	MIC_{50}	MIC_{90}
PD-0305970	0.06–0.125	0.125	0.125	0.06–0.5	0.125	0.25	0.25–1	0.25	0.5
PD-0326448	0.125–0.5	0.25	0.25	0.125–1	0.25	0.5	0.25–4	1	4
Garenoxacin	0.015–2	0.06	0.06	0.03–4	2	2	0.25–32	2	4
Gatifloxacin	0.03–2	0.125	0.125	0.06–8	4	8	1–64	8	16
Levofloxacin	0.125–4	0.25	0.25	0.125–8	8	8	16–>64	32	64

4-QUINOLIZINONES (2-PYRIDONES)

A-170568.1

A-170568.1 belongs to the 2-pyridones class of type II topoisomerase inhibitors, with following chemical name: 8-(7-methylamino-5-azaspiro[2.4]hept-5-yl)-1-cyclopropyl-7-fluoro-9-methyl-4H-4-oxo-quinolizine-3-carboxylic acid.

A-170568.1

It is structurally distinguished from ABT-719 by a certain bicyclic amine, 7-methylamino-5-azaspiro[2.4]hept-5-yl, replacing the 3-aminopyrrolidin-1-yl moiety located at C-8 position of ABT-719. The 7-amino-5-azaspiro[2.4]hept-5-yl residue is found in several new quinolones such as sitafloxacin, DC-159a, DK-507k and HSR-903. A-170568.1 is more hydrophobic than ABT-719 due to the N-methyl and bicyclic amine functionalities.

In data presented by Nilius et al. [180] A-170568.1 showed superior activity as compared to ciprofloxacin, sparfloxacin, trovafloxacin, and clinafloxacin.

The efficacy of A-170568.1 against experimental infections was reported by Meulbroek, et al. A-170568.1 was less efficacious than ciprofloxacin and trovafloxacin against Gram-negative organisms in the mouse protection test [181]. However, against Gram-positive organisms, A-170568.1 displayed significantly better efficacy than both ciprofloxacin and vancomycin. The ED$_{50}$ against ciprofloxacin-resistant MRSA in rat was at 27.1 mg/kg/day, 2-fold higher than that of vancomycin.

ABT-719

ABT-719, 8-[(3S)-aminopyrrolidin-1-yl]-1-cyclopropyl-7-fluoro-9-methyl-4H-4-oxo-quinolizine-3-carboxylic acid hydrochloride monohydrate, belong to the 2-pyridones class of type II topoisomerase inhibitors. ABT-719 was also referred to as A-86719.1 for the hydrochloride salt and as A-86719.0 for the free base [182].

ABT-719

The hydrochloride salts of ABT-719 was found to be extremely soluble in water (>16 mg/mL). Beside the 2-pyridones structure, the distinctive structural features of ABT-719 are a 3-aminopyrrolidinyl moiety at the C-8 position, as found in tosufloxacin and clinafloxacin, and a methyl group at the C-9 position. Also, a cyclopropyl group has been preserved at position 1 similar to ciprofloxacin.

Antibacterial activity of ABT-719 in comparison to the activities of ciprofloxacin, sparfloxacin and clinafloxacin, were reported by Flamm et al. [183]. In general, against Gram-positive bacteria, ABT-719 showed superior activity as compared to ciprofloxacin, sparfloxacin, trovafloxacin, and clinafloxacin. ABT-719 was found to be particularly active against resistant organisms. MIC_{90} values of ABT-719 against ciprofloxacin-resistant *S. aureus* were 0.25 and 0.5 μg/mL, respectively, which were 2- to 4-fold better than that of trovafloxacin and up to 512-fold better than that of ciprofloxacin. MIC_{90} of ABT-719 against MRSA (1 μg/mL) was identical to that of clinafloxacin, and significantly lower than those of ciprofloxacin and sparfloxacin (Table 21).

Chin et al. have also published a study of ABT-719 against 369 Gram-positive clinical isolates including MSSA and MRSA [184]. The in vitro activity of ABT-719 was compared with that of ciprofloxacin, sparfloxacin, and sitafloxacin. ABT-719 was found to be superior to ciprofloxacin and sparfloxacin, and comparable to or more potent than sitafloxacin. ABT-719 exhibited MIC_{90}s of ≤0.12 μg/mL against both ciprofloxacin-susceptible MSSA and MRSA, and ≤1 μg/mL against ciprofloxacin-resistant MRSA.

Table 21. Comparative antibacterial activities (MICs, μg/mL) of ABT-719, ciprofloxacin, clinafloxacin, and sparfloxacin against *S. aureus*

Compound	Organism (number)					
	Ciprofloxacin-susceptible MSSA (21) [Ciprofloxacin-resistant MSSA (25)]			MRSA (18)		
	Range	MIC_{50}	MIC_{90}	Range	MIC_{50}	MIC_{90}
ABT-719	0.008–0.015 [≤0.03–2]	0.008 [0.25]	0.015 [0.25]	0.06–1	0.25	1.0
Ciprofloxacin	0.12–0.5 [8–>128]	0.25 [16]	0.5 [64]	8–>128	16	>128
Clinafloxacin	0.015–0.03 [0.25–2]	0.03 [0.5]	0.03 [1]	0.25–2.0	0.25	1.0
Sparfloxacin	0.03–0.06 [0.12–16]	0.03 [8]	0.06 [16]	0.5–16	4	8

Alder et al. published several study of ABT-719 in systemic infections caused by important pathogens including staphylococci. ABT-719 displayed superior efficacy over both ciprofloxacin and vancomycin against ciprofloxacin-sensitive *S. aureus* with ED_{50}s <3.1 mg/kg/day, compared to 8-18 mg/kg/day for ciprofloxacin and vancomycin. ABT-719 was effective against ciprofloxacin-resistant MRSA with an ED_{50} value of 50.0 mg/kg/day, which was at least 10 times lower than that of ciprofloxacin. However, ABT-719 was 7-fold less efficacious than vancomycin against MRSA in these in vivo studies, despite being nearly 8-fold lower in MIC [182].

CBR-2092

In a program pursued by Cumbre Pharmaceuticals Inc. toward the rifamycin-based hybrids, a series of compounds was prepared in which rifamycin and a quinolone pharmacophore were covalently connected. Among several hybrids, CBR-2092 has been developed to Phase I clinical trials for infections involving biofilms with indwelling medical devices (catheters, prosthetic devices).

In the CBR-2092 molecule, the rifamycin SV pharmacophore is combined with a quinolone pharmacophore derived from ABT-719 [185]. ABT-719 is the most extensively characterized member of the 2-pyridone series and exhibits equipotent inhibitory activity against ciprofloxacin-resistant isolates of *S. aureus* and MRSA [186]. Rifamycins are a RNA polymerase inhibitor that have proven efficacy in the treatment of persistent bacterial infections [187, 188]. As all of the pharmacophoric features of the rifamycin identified as critical elements in the RNA polymerase interaction are preserved in CBR-2092 and the quinolone scaffold is appended to the C-3 position via hydrazone functionality, it is perhaps not surprising that CBR-2092 exhibits in vitro potency as an RNA polymerase inhibitor that is nearly equivalent to that of rifampin. Thus, in biochemical studies, CBR-2092 exhibited rifampin-like potency as an inhibitor of RNA polymerase and was an equipotent inhibitor of DNA gyrase and DNA topoisomerase IV [185].

CBR-2092

CBR-2092 exhibits potent activity against prevalent staphylococci and streptococci; especially, against the MRSA. For the MRSA isolates, the MIC ranges determined for CBR-2092 and rifampin were ≤0.004 to 2 and ≤0.004 to >4 µg/ml, respectively (Table 22).

Table 22. Comparative antibacterial activities (MICs, µg/mL) of CBR-2092, rifampin and ciprofloxacin against *S. aureus*

Compound	Organism (number)					
	MSSA (51)			MRSA (54)		
	Range	MIC_{50}	MIC_{90}	Range	MIC_{50}	MIC_{90}
CBR-2092	≤0.004–0.03	0.008	0.015	≤0.004–2	0.015	0.015
Rifampin	≤0.008–0.06	0.015	0.015	≤0.008–>4	0.015	0.015
Ciprofloxacin	0.06–>4	0.25	1	0.12–>4	>4	>4

In studies of the intracellular killing of *S. aureus*, CBR-2092 exhibited prolonged bactericidal activity that was superior to the activities of moxifloxacin, rifampin, and a combination of moxifloxacin and rifampin [189].

SUMMARY

Progressive efforts have been made to optimize the biological activity of the quinolones against *S. aureus* and drug-resistant strains including MRSA [190-194]. Much of the improved potency of modern quinolones against staphylococci has been achieved by tinkering with the N-1, C-7, and C-8 substituents on the quinolone ring system. Even by development of 6-desfluoroquinolones, it demonstrated that the spatial arrangements with appropriate substituents at N-1, C-7, and C-8 could diminish the impact of 6-fluoro- substituent.

As mentioned above, the N-1 cyclopropyl group, which was originally described for ciprofloxacin, remains one of the most effective functionalities for providing anti-MRSA quinolones. Combination of the N-1 cyclopropyl or 2-fluorocyclopropyl with a C-8 substituent (halogen, methyl, methoxy or difluoromethoxy) can improve activity of quinolones against MRSA. Unusual SAR was found in the case of gemifloxacin, zabofloxacin and DW286, where the naphthyridone nucleus provided MICs improved over those of the corresponding C-8 quinolones. Introduction of an aminopyrrolidinyl substituent at C-7 generally results in increased anti-MRSA activity compared to the piperazine derivative. Increasing the lipophilicity of a quinolone, however, as has been effected in many new agents, generally tends to increase potency against MRSA while somewhat attenuating Gram-negative potency.

Several newly developed quinolones and related compounds including 6-fluoroquinolones (delafloxacin, olamufloxacin, Dk-507k, DC-159a, DQ-113 and WCK 771), 6-fluoronaphthyridones (gemifloxacin, zabofloxacin and DW286), 6-desfluoroquinolones (garenoxacin, nemonoxacin, ozenoxacin and DX-619), quinazolinediones (PD 0305970 and PD 0326448) and 4-quinolizinones (ATB-719 and CBR-2092) exhibit enhanced activity against *S. aureus* and have proved useful against MRSA strains and are under rapid preclinical and clinical development.

REFERENCES

[1] Deresinski, S. Methicillin-resistant *Staphylococcus aureus:* an evolutionary, epidemiologic, and therapeutic odyssey. *Clin Infect Dis* 2005; 40: 562–573.
[2] Shittu, A; Lin, J. Newer antistaphylococcal agents: in-vitro studies and emerging trends in *Staphylococcus aureus* resistance. *Wounds*, 2006; 18: 129–146.
[3] Lescher, GY; Froelich, EJ; Gruett, MD; Bailey, JH; Brundage, RP. 1,8-Naphthyridine derivatives: a new class of chemotherapy agents. *J Med Pharm Chem* 1962; 5: 1063–1068.
[4] Wise R. Norfloxacin – a review of pharmacology and tissue penetration. *J Antimicrob Chemother* 1984; 13 (Suppl. B): 59–64.
[5] Appelbaum, PC; Hunter, PA. The fluoroquinolone antibacterials: past, present and future perspectives. *Int J Antimicrob Agents* 2000; 16: 5–15.
[6] Koga, H; Itoh, A; Murayama, S; Suzue, S; Irikura, T. Structure-activity relationships of antibacterial 6,7- and 7,8-disubstituted 1-alkyl-1,4-dihydro-4-oxoquinoline-3-carboxylic acids. *J Med Chem* 1980; 23: 1358–63.
[7] Asahina, Y; Ishizaki, T; Suzue, S. Recent advances in structure activity relationships in new quinolones. *Prog Drug Res* 1992; 38: 57–106.
[8] Andriole, VT. *The Quinolones*, Academic Press: London, 1988.
[9] Wolfson, J.S.; Hooper, D.C. *Quinolone Antimicrobial Agents*, American Society for Microbilogy: Washington DC, 1989.
[10] Domagala JM. Structure-activity and structure side effect relationships for the quinolone antibacterials. *J Antimicrob Chemother* 1994; 33: 685–706.
[11] Ball, P; Fernald, A; Tillotson, G. Therapeutic advances of new fluoroquinolones. *Expert Opin Inv Drugs* 1998; 7: 761–783.

[12] Zhanel, GG; Ennis, K; Vercaigne, L; Walkty, A; Gin, AS; Embil, J; Smith, H; Hoban, DJ. A critical review of the fluoroquinolones: focus on respiratory infections. *Drugs* 2002; 62: 13–59.

[13] De Sarro, A; De Sarro, G. Adverse reactions to fluoroquinolones. an overview on mechanistic aspects. *Curr Med Chem* 2001; 8: 371–384.

[14] Hooper DC. Mechanisms of fluoroquinolone resistance. *Drug Resis Updates* 1999; 2: 38–55.

[15] Hooper DC. Bacterial topoisomerases, anti-topoisomerases, and antitopoisomerase resistance. *Clin Infect Dis* 1998; 27(Suppl 1): S54–63.

[16] Schmitz, F-J; Higgins, PG; Mayer, S; Fluit, AC; Dalhoff, A. Activity of quinolones against Gram-positive cocci: mechanisms of drug action and bacterial resistance. *Eur J Clin Microbiol Infect Dis* 2002; 21: 647–659.

[17] Wang, JC; Lynch, AS. Transcription and DNA supercoiling. *Curr Opin Genet Dev* 1993; 3: 764–768.

[18] Levine, C; Hiasa, H; Marians, KJ. DNA gyrase and topoisomerase IV: Biochemical activities, physiological roles during chromosome replication, and drug sensitivites. *Biochim Biophys Acta Gene Struct Expression* 1998; 1400: 29–43.

[19] Hawkey, PM. Mechanisms of quinolone action and microbial response. *J Antimicrob Chemother* 2003; 51(Suppl. S1): 29–35.

[20] Drlica, K. Control of bacterial DNA supercoiling. *Mol Microbiol* 1992; 6: 425–433.

[21] Hooper DC. Fluoroquinolone resistance among Gram-positive cocci. *Lancet Infect Dis* 2002; 2: 530–538.

[22] Wigley, DB. Structure and mechanism of DNA topoisomerases. *Ann Rev Biophys Biomol Struct* 1995; 24: 185–208.

[23] Drlica, K; Zhao, X. DNA gyrase, topoisomerase IV, and the 4-quinolones. *Microbiol Mol Biol Rev* 1997; 61: 377–392.

[24] Shen, LL; Mitscher, LA; Sharma, PN; O'Donnell, TJ; Chu, WT; Cooper, CS; Rosen, T; Pernet, AG. Mechanism of inhibition of DNA gyrase by quinolone antibacterials: a co-operative drug-DNA binding model. *Biochemistry*, 1989; 28: 3886–3894.

[25] Van Bambeke, F; Michot, J-M; Van Eldere, J; Tulkens, PM. Quinolones in 2005: an update. *Clin Microbiol Infect* 2005; 11: 256–280.

[26] Anderson, VE; Zaniewski, RP; Kaczmarek, FS; Gootz, TD; Osheroff, N. Quinolones inhibit DNA religation mediated by *Staphylococcus aureus* topoisomerase IV. Changes in drug mechanism across evolutionary boundaries. *J Biol Chem* 1999; 274: 35927–35932.

[27] Gootz, TD; Zaniewski, RP; Haskell, SL; Kaczmarek, FS; Maurice, AE. Activities of trovafloxacin compared with those of other

fluoroquinolones against purified topoisomerases and *gyrA* and *grlA* mutants of *Staphylococcus aureus*. *Antimiocrob Agents Chemother* 1999; 43: 1845–1855.
[28] Ruiz, J; Sierra, JM; de Anta, MT; Vila, J. Characterization of sparfloxacin-resistant mutants of *Staphylococcus aureus* obtained in vitro. *Int J Antimicrob Agents* 2001; 18: 107–112.
[29] Hooper, DC. Mode of action of fluoroquinolones. *Drugs* 1999; 58 (Suppl. 2): 6–10.
[30] Pan, XS; Fisher, LM. DNA gyrase and topoisomerase IV are dual targets of clinafloxacin action in *Streptococcus pneumoniae*. *Antimicrob Agents Chemother* 1998; 42: 2810–2816.
[31] Heaton, VJ; Ambler, JE; Fisher, LM. Potent antipneumococcal activity of gemifloxacin is associated with dual targeting of gyrase and topoisomerase IV, an in vivo target preference for gyrase, and enhanced stabilization of cleavable complexes in vitro. *Antimicrob Agents Chemother* 2000; 44: 3112–3117.
[32] Nilius, AM; Shen, LL; Hensey-Rudloff, D; Almer, LS; Beyer, JM; Balli, DJ; Cai, Y; Flamm, RK. In vitro antibacterial potency and spectrum of ABT-492, a new fluoroquinolone. *Antimicrob Agents Chemother* 2003; 47: 3260–3269.
[33] Roychoudhury, S; Ledoussal, B. Non-fluorinated quinolones (NFQs): new antibacterials with unique properties against quinolone-resistant Gram-positive pathogens. *Curr Drug Targets - Infectious Disorders* 2002; 2: 51–65.
[34] Takei, M; Fukuda, H; Kishii, R; Hosaka, M. Target preference of 15 quinolones against *Staphylococcus aureus*, based on antibacterial activities and target inhibition. *Antimicrob Agents Chemother* 2001; 45: 3544–3547.
[35] Schulte, A; Heisig, P. In vitro activity of gemifloxacin and five other fluoroquinolones against defined isogenic mutant of *Escherichia coli*, *Pseudomonas aeruginosa* and *Staphylococcus aureus*. *J Antimicrob Chemother* 2000; 46: 1037–1046.
[36] Blanche, F.; Cameron, B; Bernard, F-X; Maton, L; Manse, B; Ferrero, L; Ratet, N; Lecoq, C; Goniot, A; Bisch, D; Crouzet, J. Differential behaviors of *Staphylococcus aureus* and *Escherichia coli* type II DNA topoisomerases. *Antimicrob Agents Chemother* 1996; 40: 2714–2720.
[37] Tanaka, M; Onodera, Y; Uchida, Y; Sato, K; Hayakawa, I. Inhibitory activities of quinolones against DNA gyrase and topoisomerase IV purified from *Staphylococcus aureus*. *Antimicrob Agents Chemother* 1997; 41: 2362–2366.

[38] Schmitz, F-J; Boos, M; Mayer, S; Jagusch, H; Fluit, AC. Increased in vitro activity of the novel des-fluoro(6) quinolone BMS-284756 against genetically defined clinical isolates of *Staphylococcus aureus*. *J Antimicrob Chemother* 2002; 49: 283–287.

[39] Blumberg, HM; Rimland, D; Carroll, DJ; Terry, P; Wachsmuth, IK. Rapid development of ciprofloxacin resistance in methicillin-susceptible and -resistant *Staphylococcus aureus*. *J Infect Dis* 1991; 163: 1279–1285.

[40] Harnett, N; Brown, S; Krishnan, C. Emergence of quinolone resistance among clinical isolates of methicillin-resistant *Staphylococcus aureus* in Ontario, Canada. *Antimicrob Agents Chemother* 1991; 35: 1911–1913.

[41] Barry, AL; Jones, RN. In vitro activity of ciprofloxacin against gram-positive cocci. *Am J Med* 1987; 82: 27–32.

[42] Scheel, O; Lyon, DJ; Rosdahl, VT; Adeyemi-Doro, FA; Ling, TK; Cheng, AF. In-vitro susceptibility of isolates of methicillin-resistant *Staphylococcus aureus* 1988-1993. *J Antimicrob Chemother* 1996; 37: 243–251.

[43] Schmitz, FJ; Fluit, AC; Hafner, D; Beeck, A, Perdikouli, M; Boos, M; Scheuring, S; Verhoef, J; Köhrer, K; Von Eiff, C. Development of resistance to ciprofloxacin, rifampin, and mupirocin in methicillin-susceptible and –resistant *Staphylococcus aureus* isolates. *Antimicrob Agents Chemother* 2000; 44: 3229–3231.

[44] Lipsitch, M; Samore, MH. Antimicrobial use and antimicrobial resistance: a population perspective. *Emerg Infect Dis* 2002; 8: 347–354.

[45] Yoshida, H; Bogaki, M; Nakamura, S; Ubukata, K; Konno, M. Nucleotide sequence and characterization of the *Staphylococcus aureus* norA gene, which confers resistance to quinolones. *J Bacteriol* 1990; 172: 6942–6949.

[46] Goswitz, JJ; Willard, KE; Fasching, CE; Peterson, LR. Detection of *gyrA* mutations associated with ciprofloxacin resistance in methicillin-resistant *Staphylococcus aureus*: analysis by polymerase chain reaction and automated direct DNA sequencing. *Antimicrob Agents Chemother* 1992; 36: 1166–1169.

[47] Ferrero, L; Cameron, B; Manse, B; Lagneaux, D; Crouzet, J; Famechon, A; Blanche, F. Cloning and primary structure of *Staphylococcus aureus* DNA topoisomerase IV: a primary target of fluoroquinolones. *Mol Microbiol* 1994; 13: 641–653.

[48] Willmott, CJ; Maxwell, A. A single point mutation in the DNA gyrase A protein greatly reduces binding of fluoroquinolones to the gyrase-DNA complex. *Antimicrob Agents Chemother* 1993; 37: 126–127.

[49] Fournier, B; Hooper, DC. Mutations in topoisomerase IV and DNA gyrase of *Staphylococcus aureus*: Novel pleiotropic effects on quinolone and coumarin activity. *Antimicrob Agents Chemother* 1998; 42: 121–128.

[50] Ng, EY; Trucksis, M; Hooper, DC. Quinolone resistance mutations in topoisomerase IV: relationship of the *flqA* locus and genetic evidence that topoisomerase IV is the primary target and DNA gyrase the secondary target of fluoroquinolones in *Staphylococcus aureus*. *Antimicrob Agents Chemother* 1996; 40: 1881–1888.

[51] Schmitz, FJ; Hofmann, B; Hansen, B; Scheuring, S; Lückefahr, M; Klootwijk, M; Verhoef, J; Fluit, A; Heinz, HP; Köhrer, K; Jones, ME. Relationship between ciprofloxacin, ofloxacin, levofloxacin, sparfloxacin and moxifloxacin (BAY 12-8039) MICs and mutations in *grlA*, *grlB*, *gyrA* and *gyrB* in 116 unrelated clinical isolates of *Staphylococcus aureus*. *J Antimicrob Chemother* 1998; 41: 481–484.

[52] Wang, T; Tanaka, M; Sato, K. Detection of *grlA* and *gyrA* mutations in 344 *Staphylococcus aureus* strains. *Antimicrob Agents Chemother* 1998; 42: 236–240.

[53] Shreedhan, S; Oram, M; Jensen, B; Peterson, LR; Fisher, LM. DNA gyrase *gyrA* mutations in ciprofloxacin-resistant strains of *Staphylococcus aureus*: close similarities with quinolone resistance mutations in *Escherichia coli*. *J Bacteriol* 1990; 172: 7260–7262.

[54] Bast, DJ; Low, DE; Duncan, CL; Kilburn, L; Mandell, LA; Davidson, RJ; de Azavedo, JC. Fluoroquinolone resistance in clinical isolates of *Streptococcus pneumoniae*: Contributions of type II topoisomerase mutations and efflux to levels of resistance. *Antimicrob Agents Chemother* 2000; 44: 3049–3054.

[55] Piddock, LJV; Johnson, MM; Simjee, S; Pumbwe, L. Expression of efflux pump gene *pmrA* in fluoroquinolone-resistant and -susceptible clinical isolates of *Streptococcus pneumoniae*. *Antimicrob Agents Chemother* 2002; 46: 808–812.

[56] Ince, D; Zhang, X; Silver, LC; Hooper, DC. Dual targeting of DNA gyrase and topoisomerase IV: target interactions of garenoxacin (BMS-284756, T-3811ME), a new desfluoroquinolone. *Antimicrob Agents Chemother* 2002; 46: 3370–3380.

[57] Bryskier, A; Chantot, J-F. Classification and structure-activity relationships of fluoroquinolones. *Drugs* 1995; 49 (Suppl. 2): 16–28.

[58] Peterson, LR. Quinolone molecular structure-activity relationships: what we have learned about improving antimicrobial activity. *Clin Infect Dis* 2001; 33 (Suppl 3): S180–S186.

[59] Roychoudhury, S; Twinem, TL; Makin, KM; McIntosh, EJ; Ledoussal, B; Catrenich, CE. Activity of non-fluorinated quinolones (NFQs) against quinolone-resistant *Escherichia coli* and *Streptococcus pneumoniae*. *J Antimicrob Chemother* 2001; 48: 29–36.

[60] Hartman-Neumann, S; DenBleyker, K; Pelose, LA; Lawrence, LE; Barett, JF; Douherty, TJ. Selection and genetic characterization of *Streptococcus pneumoniae* mutants resistant to the des-F(6) quinolone BMS-284756. *Antimicrob Agents Chemother* 2001; 45: 2865–2870.

[61] Fukuda, H; Kishii, R; Takei, M; Hosaka, M. Contributions of the 8-methoxy group of gatifloxacin to resistance selectivity, target preference, and antibacterial activity against *Streptococcus pneumoniae*. *Antimicrob Agents Chemother* 2001; 45: 1649–1653.

[62] Ince, D; Hooper, DC. Mechanisms and frequency of resistance to gatifloxacin in comparison to AM-1121 and ciprofloxacin in *Staphylococcus aureus*. *Antimicrob Agents Chemother* 2001; 45: 2755–2764.

[63] Zhao, X; Wang, J-Y; Xu, C; Dong, Y; Zhou, J; Domagala, J; Drlica, K. Killing of *Staphylococcus aureus* by C-8-methoxy fluoroquinolones. *Antimicrob Agents Chemother* 1998; 42: 956–958.

[64] Zhanel, GG; Hoban, DJ; Schurek, K; Karlowsky, JA. Role of efflux mechanisms on fluoroquinolone resistance in *Streptococcus pneumoniae* and *Pseudomonas aeruginosa*. *Int J Antimicrob Agents* 2004; 24: 529–535.

[65] Markham, PN; Neyfakh, AA. Inhibition of the multidrug transporter NorA prevents emergence of norfloxacin resistance in *Staphylococcus aureus*. *Antimicrob Agents Chemother* 1996; 40: 2673–2674.

[66] Markham, PN. Inhibition of the emergence of ciprofloxacin resistance in *Streptococcus pneumoniae* by the multidrug efflux inhibitor reserpine. *Antimicrob Agents Chemother* 1999; 43: 988–989.

[67] Sulavik, CM; Barg, NL. Examination of methicillin-resistant and methicillin-susceptible *Staphylococcus aureus* mutants with low-level fluoroquinolone resistance. *Antimicrob Agents Chemother* 1998; 42: 3317–3319.

[68] Scheld, WM. Maintaining fluoroquinolone class efficacy: review of influencing factors. *Emerg Infect Dis* 2003; 9: 1–9.

[69] Beyer, R; Pestova, E; Millichap, JJ; Stosor, V; Noskin, GA; Peterson, LR. A convenient assay for estimating the possible involvement of efflux of fluoroquinolones by *Streptococcus pneumoniae* and *Staphylocccus aureus*: evidence for diminished moxifloxacin, sparfloxacin, and trovafloxacin efflux. *Antimicrob Agents Chemother* 2000; 44: 798–801.

[70] Aeschlimann, JR; Dresser, LD; Kaatz, GW; Rybak, MJ. Effects of NorA inhibitors on *in vitro* antibacterial activities and postantibiotic effects of levofloxacin, ciprofloxacin, and norfloxacin in genetically related strains of *Staphylococcus aureus*. *Antimicrob Agents Chemother* 1999; 43: 335–340.

[71] Piddock, LJ; Johnson, MM. Accumulation of 10 fluoroquinolones by wild-type or efflux mutant *Streptococcus pneumoniae*. *Antimicrob Agents Chemother* 2002; 46: 813–820.

[72] Brenwald, NP; Gill, MJ; Wise, R. Prevalence of a putative efflux mechanism among fluoroquinolone-resistant clinical isolates of *Streptococcus pneumoniae*. *Antimicrob Agents Chemother* 1998; 42: 2032–2035.

[73] Zeller, V; Janoir, C; Kitzis, MD; Gutmann, L; Moreau, NJ. Active efflux as a mechanism of resistance to ciprofloxacin in *Streptococcus pneumoniae*. *Antimicrob Agents Chemother* 1997; 41: 1973–1978.

[74] Takenouchi, T; Tabata, F; Iwata, Y; Hanzawa, H; Sugawara, M; Ohya, S. Hydrophilicity of quinolones is not an exclusive factor for decreased activity in efflux-mediated resistant mutants of *Staphylococcus aureus*. *Antimicrob Agents Chemother* 1996; 40: 1835–1842.

[75] Madras-Kelly, KJ; Daniels, C; Haegbloom, M; Thompson, M. Pharmacodynamic characterization of efflux and topoisomerase IV-mediated fluoroquinolone resistance in *Streptococcus pneumoniae*. *J Antimicrob Chemother* 2002; 50: 211–218.

[76] Kerns, RJ; Rybak, MJ; Kaatz, GW; Vaka, F; Cha, R; Grucz, RG; Diwadkar, VU; Ward, TD. Piperazinyl-linked fluoroquinolone dimers possessing potent antibacterial activity against drug-resistant strains of *Staphylococcus aureus*. *Bioorg Med Chem Lett* 2003; 13: 1745–1749.

[77] Kerns, RJ; Rybak, MJ; Kaatz, GW; Vaka, F; Cha, R; Grucz, RG; Diwadkar, VU. Structural features of piperazinyl-linked ciprofloxacin dimers required for activity against drug-resistant strains of *Staphylococcus aureus*. *Bioorg Med Chem Lett* 2003; 13: 2109–2112.

[78] Bhanot, SK; Singh, M; Chatterjee, NR. The chemical and biological aspects of fluoroquinolones: reality and dreams. *Curr Pharm Design* 2001; 7: 311–335.

[79] De Sarro, A; De Sarro, G. Adverse reactions to fluoroquinolones. an overview on mechanistic aspects. *Curr Med Chem* 2001; 8: 371–384.

[80] Takemura, M; Hayakawa, I. Recent advances in the field of quinolones. *Il Farmaco* 2001; 56: 37–40.

[81] Kimura, Y; Atarashi, S; Kawakami, K; Sato, K; Hayakawa, I. (Fluorocyclopropyl)quinolones. 2. Synthesis and Stereochemical structure-activity relationships of chiral 7-(7-amino-5-

azaspiro[2.4]heptan-5-yl)-1-(2-fluorocyclopropyl)quinolone antibacterial agents. *J Med Chem* 1994; 37: 3344–3352.

[82] Brighty, KE; Gootz, TD. The chemistry and biological profile of trovafloxacin. *J Antimicrob Chemother* 1997; 39 (Suppl. B): 1–14.

[83] Chung, SJ; Kim, DH. Synthesis and evaluation of 3-fluoro-2-piperazinyl-5,8,13-trihydro-5-oxoquino[1,2-a][3,1]benzoxazine-6-carboxylic acids as potential antibacterial agents. *Arch Pharm* 1997; 330: 63–66.

[84] Morrissey, I; Hoshino, K; Sato, K; Yoshida, A; Hayakawa, I; Bures, MG; Shen, LL. Mechanism of differential activities of ofloxacin enantiomers. *Antimicrob Agents Chemother* 1996; 40: 1775–1784.

[85] Tillotson, GS; Blondeau, JM. Structure–activity–function evaluation of the fluoroquinolones. In: Adam D, Finch RG and Hunter PA (eds.) *Moxifloxacin in Practice*. Maxim Medical (1999) 91–101.

[86] Tillotson, GS. Quinolones: structure–activity relationships and future predictions. *J Med Microbiol* 1996; 44: 320–324.

[87] Segawa, J; Kitano, M; Kazuno, K; Matsuoka, M; Shirahase, I; Ozaki, M; Matsuda, M; Tomii, Y; Kise, M. Studies on pyridonecarboxylic acids. 1. Synthesis and antibacterial evaluation of 7-substituted-6-halo-4-oxo-4H-[1,3]thiazeto[3,2-a]quinoline-3- carboxylic acids. *J Med Chem* 1992; 35: 4727–4738.

[88] Chu, DT; Hallas, R; Clement, JJ; Alder, J; McDonald, E; Plattner, JJ. Synthesis and antitumour activities of quinolone antineoplastic agents. *Drugs Exp Clin Res* 1992; 18: 275–282.

[89] Yoshida, T; Yamamoto, Y; Orita, H; Kakiuchi, M; Takahashi, Y; Itakura, M; Kado, N; Mitani, K; Yasuda, S; Kato, H; Itoh, Y. Studies on quinolone antibacterials. IV. Structure-activity relationships of antibacterial activity and side effects for 5- or 8-substituted and 5,8-disubstituted-7-(3-amino-1-pyrrolidinyl)-1-cyclopropyl-1, 4-dihydro-4-oxoquinoline-3-carboxylic acids. *Chem Pharm Bull (Tokyo)* 1996; 44: 1074–1085.

[90] Jaillon, P; Morganroth, J; Brumpt, I; Talbot, G. Overview of electrocardiographic and cardiovascular safety data for sparfloxacin. Sparfloxacin Safety Group. *J Antimicrob Chemother* 1996; 37 (Suppl. A): 161–167.

[91] Schentag, JJ. Sparfloxacin: a review. *Clin Ther* 2000; 22: 372–378.

[92] Wiedemann, B; Heisig, P. Antibacterial activity of grepafloxacin. *J Antimicrob Chemother* 1997; 40 (Suppl. A): 19–25.

[93] Bryskier, A; Chantot, J-F. Classification and structure-activity relationships of fluoroquinolones. *Drugs* 1995; 49 (Suppl. 2): 16–28.

[94] Gootz, TD; Brighty, KE. Fluoroquinolone antibacterials: SAR mechanism of action, resistance, and clinical aspects. *Med Res Rev* 1996; 16: 433–486.
[95] Lawrence, LE; Wu, P; Fan, L; Gouveia, KE; Card, A; Casperson, M; Denbleyker, K; Barrett, JF. The inhibition and selectivity of bacterial topoisomerases by BMS-284756 and its analogues. *J Antimicrob Chemother* 2001; 48, 195–201.
[96] Barry, AL; Fuchs, PC; Brown, SD. In vitro activities of three nonfluorinated quinolones against representative bacterial isolates. *Antimicrob Agents Chemother* 2001; 45: 1923–1927.
[97] Cecchetti, V; Clementi, S; Cruciani, G; Fravolini, A; Pagella, PG; Savino, A; Tabarrini, O. 6-Aminoquinolones: A new class of quinolone antibacterials? *J Med Chem* 1995; 38: 973–982.
[98] Cecchetti, V; Fravolini, A; Lorenzini, MC; Tabarrini, O; Terni, P; Xin, T. Studies on 6-aminoquinolones: Synthesis and antibacterial evaluation of 6-amino-8-methyl quinolones. *J Med Chem* 1996; 39: 436–445.
[99] Cecchetti, V; Fravolini, A; Palumbo, M; Sissi, C; Tabarrini, O; Terni, P; Xin, T. Potent 6-desfluoro-8-methylquinolones as new lead compounds in antibacterial chemotherapy. *J Med Chem* 1996; 39: 4952–4957.
[100] Artico, M; Mai, A; Spardella, G; Massa, S; Musiu, C; Lostia, S; Demontis, F; La Colla; F. Nitroquinolones with broad-spectrum antimycobacterial activity in vitro. *Bioorg Med Chem Lett* 1999; 9: 1651–1656.
[101] Gootz, TD; Brighty, KE. *The Quinolones*. 2nd ed., Academic Press, 1998.
[102] Sun, J; Sakai, S; Tauchi, Y; Deguchi, Y; Chen, J; Zhang, R; Morimoto, K. Determination of lipophilicity of two quinolone antibacterials, ciprofloxacin and grepafloxacin, in the protonation equilibrium. *Eur J Pharm Biopharm* 2002; 54: 51–58.
[103] Ball, P. Moxifloxacin (Avelox): an 8-methoxyquinolone antibacterial with enhanced potency. *Int J Clin Pract* 2000; 54: 329–332.
[104] Kim, JH; Kang, JA; Kim, YG; Kim, JW; Lee, JH; Choi, EC; Kim, BK. In vitro and in vivo antibacterial efficacies of CFC-222, a new fluoroquinolone. *Antimicrob Agents Chemother* 1997; 41: 2209–2213.
[105] Weller, TMA; Andrews, JM; Jevons, G; Wise, R. The in vitro activity of BMS-284756, a new des-fluorinated quinolone. *J Antimicrob Chemother* 2002; 49: 177–184.
[106] Foroumadi, A; Emami, S; Davood, A; Moshafi, MH; Sharifian, A; Tabatabaiee, M; Tarhimi Farimani, H; Sepehri, G; Shafiee, A. Synthesis and in-vitro antibacterial activities of N-substituted piperazinyl quinolones. *Pharm Sci* 1997; 3: 559.

[107] Piddock, LJV; Jin, YF; Griggs, DJ. Effect of hydrophobicity and molecular mass on the accumulation of fluoroquinolones by *Staphylococcus aureus*. *J Antimicrob Chemother* 2001; 47: 261–270.
[108] McCaffrey, C; Bertasso, A; Pace, J; Georgopapadakou, NH. Quinolone accumulation in *Escherichia coli, Pseudomonas aeruginosa*, and *Staphylococcus aureus*. *Antimicrob Agents Chemother* 1992; 36: 1601–1605.
[109] Piddock, LJV; Hall, MC; Wise, R. Mechanism of action of lomefloxacin. *Antimicrob Agents Chemother* 1990; 34: 1088–1093.
[110] Foroumadi, A; Emami, S; Mehni, M; Moshafi, MH; Shafiee, A. Synthesis and antibacterial activity of N-[2-(5-bromothiophen-2-yl)-2-oxoethyl] and N-[(2-5-bromothiophen-2-yl)-2-oximinoethyl] derivatives of piperazinyl quinolones. *Bioorg Med Chem Lett* 2005; 15: 4536–4539.
[111] Foroumadi, A; Oboudiat, M; Emami, S; Karimollah, A; Saghaee, L; Moshafi, MH; Shafiee, A. Synthesis and antibacterial activity of *N*-[2-[5-(methylthio)thiophen-2-yl]-2-oxoethyl] and *N*-[2-[5-(methylthio)thiophen-2-yl]-2-(oxyimino)ethyl] piperazinylquinolone derivatives. *Bioorg Med Chem* 2006; 14: 3421–3427.
[112] Foroumadi, A; Ghodsi, S; Emami, S; Najjari, S; Samadi, N; Faramarzi, MA; Beikmohammadi, L; Shirazi, FH; Shafiee, A. Synthesis and antibacterial activity of new fluoroquinolones containing a substituted *N*-(phenethyl)piperazine moiety. *Bioorg Med Chem Lett* 2006; 16: 3499–3503.
[113] Foroumadi, A; Mohammadhosseini, N; Emami, S; Letafat, B; Faramarzi, MA; Samadi, N; Shafiee, A. Synthesis and antibacterial activity of new 7-piperazinylquinolones containing functionalized 2-(furan-3-yl)ethyl moiety. *Arch Pharm* 2007; 340: 47–52.
[114] Letafat, B; Emami, S; Mohammadhosseini, N; Faramarzi, MA; Samadi, N; Shafiee, A; Foroumadi, A. Synthesis and antibacterial activity of new *N*-[2-(thiophen-3-yl)ethyl] piperazinyl quinolones. *Chem Pharm Bull* 2007; 55: 894–898.
[115] Foroumadi, A; Emami, S; Mansouri, S; Javidnia, A; Saeid-Adeli, N; Shirazi, FH; Shafiee, A. Synthesis and antibacterial activity of levofloxacin derivatives with certain bulky residues on piperazine ring. *Eur J Med Chem* 2007; 42: 985–992.
[116] Emami, S; Foroumadi, A; Faramarzi, MA; Samadi, N. Synthesis and antibacterial activity of quinolone-based compounds containing a coumarin moiety. *Arch Pharm Chem Life Sci* 2008; 341: 42–48.
[117] Foroumadi, A; Emami, S; Hassanzadeh, A; Rajaee, M; Sokhanvar, K; Moshafi, MH; Shafiee, A. Synthesis and antibacterial activity of N-(5-benzylthio-1,3,4-thiadiazol-2-yl) and N-(5-benzylsulfonyl-1,3,4-

thiadiazol-2-yl)piperazinyl quinolone derivatives. *Bioorg Med Chem Lett* 2005; 15: 4488–4492.

[118] Foroumadi, A; Firoozpour, L; Emami, S; Mansouri, S; Ebrahimabadi, AH; Asadipour, A; Amini, M; Saeid-Adeli, N; Shafiee, A. Synthesis and antibacterial activity of *N*-[5-(chlorobenzylthio)-1,3,4-thiadiazol-2-yl] piperazinyl quinolone derivatives. *Arch Pharm Res* 2007; 30: 138–145.

[119] Foroumadi, A; Soltani, F; Moshafi, MH; Ashraf-Askari, R. Synthesis and in vitro antibacterial activity of some *N*-(5-aryl-1,3,4-thiadiazole-2-yl)piperazinyl quinolone derivatives. *Il Farmaco* 2003; 58: 1023–1028.

[120] Foroumadi, A; Mansouri, S; Emami, S; Mirzaei, J; Sorkhi, M; Saeid-Adeli, N; Shafiee, A. Synthesis and antibacterial activity of nitroaryl thiadiazole-levofloxacin hybrids. *Arch Pharm* 2006; 339: 621–624.

[121] Jazayeri, S; Moshafi, MH; Firoozpour, L; Emami, S; Rajabalian, S; Haddad, M; Pahlavanzadeh, F; Esnaashari, M; Shafiee, A; Foroumadi; A. Synthesis and antibacterial activity of nitroaryl thiadiazole-gatifloxacin hybrids. *Eur J Med Chem* 2009; 44: 1205–1209.

[122] Foroumadi, A; Mansouri, S; Kiani, Z; Rahmani, A. Synthesis and in vitro antibacterial evaluation of N-[5-(5-nitro-2-thienyl)-1,3,4-thiadiazole-2-yl] piperazinyl quinolones. *Eur J Med Chem* 2003; 38: 851–854.

[123] Foroumadi, A; Ashraf-Askari, R; Moshafi, MH; Emami, S; Zeynali, A. Synthesis and in vitro antibacterial activity of N-[5-(5-nitro-2-furyl)-1,3,4-thiadiazol-2-yl]piperazinyl quinolone derivatives. *Pharmazie* 2003; 58: 432–433.

[124] Dong, Y; Xu, C; Zhao, X; Domagala, J; Drlica, K. Fluoroquinolone action against mycobacteria: effects of C-8 substituents on growth, survival, and resistance. *Antimicrob Agents Chemother* 1998; 42: 2978–2984.

[125] Lu, T; Zhao, X; Drlica, K. Gatifloxacin activity against quinolone-resistant gyrase: allele-specific enhancement of bacteriostatic and bactericidal activities by the C-8-methoxy group. *Antimicrob Agents Chemother* 1999; 43: 2969–2974.

[126] Lu, T; Zhao, X; Li, X; Drlica-Wagner, A; Wang, J-Y; Domagala, J; Drlica, K. Enhancement of fluoroquinolone activity by C-8 halogen and methoxy moieties: action against a gyrase resistance mutant of *Mycobacterium smegmatis* and a gyrase-topoisomerase IV double mutant of *Staphylococcus aureus*. *Antimicrob Agents Chemother* 2001; 45: 2703–2709.

[127] Kawakami, K, Ohki, H, Kimura, K; et al., DK-507k, a new 8-methoxyquinolone: synthesis and biological evaluation of 7-[(3-amino-

4-substituted)pyrrolidin-1-yl] derivatives. In: Abstracts of the 41st Interscience Conference on Antimicrobial Agents and Chemotherapy. 2001; Abstract F-546, p. 218.

[128] Otani, T; Tanaka, M; Ito, E; Kurosaka, Y; Murakami, Y; Onodera, K; Akasaka, T; Sato, K. In vitro and in vivo antibacterial activities of DK-507k, a novel fluoroquinolone. *Antimicrob Agents Chemother* 2003; 47: 3750–3759.

[129] Tanaka, M, Tachibana, M, Seki, H, Ohzone, Y, DK-507k, a new 8-methoxyquinolone: pharmacokinetics in rats and monkeys. In: Abstracts of the 41st Interscience Conference on Antimicrobial Agents and Chemotherapy. 2001; Abstract F-548, p. 218.

[130] Hoshino, K; Inoue, K; Murakami, Y; Kurosaka, Y; Namba, K; Kashimoto, Y; Uoyama, S; Okumura, R; Higuchi, S; Otani, T. In vitro and in vivo antibacterial activities of DC-159a, a new fluoroquinolone. *Antimicrob Agents Chemother* 2008; 52: 65–76.

[131] Jones, RN; Fritsche, TR; Sader, HS. Antimicrobial activity of DC-159a, a new fluoroquinolone, against 1,149 recently collected clinical isolates. *Antimicrob Agents Chemother* 2008; 52: 3763–3775.

[132] Okezaki, E; Watanabe, Y; Hirose, T; Yoshida, T; Aoki, Y; Kato, H. Antibacterial activity of HSR-903, a new novel quinolone, abstr. F202, p. 148. *In* Program and abstracts of the 35th Interscience Conference on Antimicrobial Agents and Chemotherapy. American Society for Microbiology, 1995; Washington, D.C.

[133] Yoshida, T; Kado, N; Okezaki, E; Yasuda, S; Kato, H. Synthesis and antibacterial activity of HSR-903 and related compounds, abstr. F58, p. 110. *In* Program and abstracts of the 36th Interscience Conference on Antimicrobial Agents and Chemotherapy. American Society for Microbiology, 1996; Washington, D.C.

[134] Takahashi, Y; Masuda, N; Otsuki, M; Miki, M; Nishino, T. In vitro activity of HSR-903, a new quinolone. *Antimicrob Agents Chemother* 1997; 41: 1326–1330.

[135] Watanabe, A; Tokue, Y; Takahashi, H; Kikuchi, T; Kobayashi, T; Gomi, K; Fujimura, S; Nukiwa, T. In vitro activity of HSR-903, a new oral quinolone, against bacteria causing respiratory infections. *Antimicrob Agents Chemother* 1999; 43: 1767–1768.

[136] Ng, EYW; Trucksis, M; Hooper, DC. Quinolone resistance mediated by *norA*: physiologic characterization and relationship to *flqB*, a quinolone resistance locus on the *Staphylococcus aureus* chromosome. *Antimicrob Agents Chemother* 1994; 37: 1345–1355.

[137] Murata, M; Takahara, E; Nagata, O; Kato, H; Tamai, I; Tsuji, A. Carrier-mediated tissue distribution and pharmacokinetics of HSR-903,

a new quinolone, abstr. F203, p. 148. *In* Program and abstracts of the 35th Interscience Conference on Antimicrobial Agents and Chemotherapy. American Society for Microbiology, 1995; Washington, D.C.

[138] Sun, J; Deguchi, Y; Tauchi, Y; He, Z; Cheng, G; Morimoto, K. Distribution characteristics of orally administered olamufloxacin, a newly synthesized fluoroquinolone antibacterial, in lung epithelial lining fluid and alveolar macrophage in rats. *Eur J Pharm Biopharm* 2006; 64: 238–245.

[139] Tanaka, M; Yamazaki, E; Chiba, M; Yoshihara, K; Akasaka, T; Takemura, M; Sato, K. In vitro antibacterial activities of DQ-113, a potent quinolone, against clinical isolates. *Antimicrob Agents Chemother* 2002; 46: 904–908.

[140] Hong, SG; Moland, ES; Wickman, PA; Black, JA; Hossain, A; Hanson, ND; Thomson, KS. In vitro studies with DQ-113 and comparison fluoroquinolones to determine propensities to select resistance in grampositive cocci. *Antimicrob Agents Chemother* 2007; 51: 1512–1514.

[141] Kaneko, Y; Yanagihara, K; Miyazaki, Y; Tsukamoto, K; Hirakata, Y; Tomono, K; Kadota, J; Tashiro, T; Murata, I; Kohno, S. Effects of DQ-113, a new quinolone, against methicillin- and vancomycin-resistant *Staphylococcus aureus*-caused hematogenous pulmonary infections in mice. *Antimicrob Agents Chemother* 2003; 47: 3694–3698.

[142] Harnett, SJ; Fraise, AP; Andrews, JM; Jevons, G; Brenwald, NP; Wise, R. Comparative study of the in vitro activity of a new fluoroquinolone, ABT-492. *J Antimicrob Chemother* 2004; 53: 783–792.

[143] Cheng, Y. Rib-X Pharmaceuticals repaints the antibioticscape. *Tech Impact @ Yale* 2008; 2: 4.

[144] Ishikawa, H; Tabusa, F; Miyamoto, H; Kano, M; Ueda, H; Tamaoka, H; Nakagawa, K. Studies on antibacterial agents. I. Synthesis of substituted 6,7-dihydro-1-oxo-1H,5H-benzo[i,j]-quinolizine-2-carboxylic acids. *Chem Pharm Bull* 1989; 37: 2103–2108.

[145] Kurokawa, I; Akamatsu, H; Nishigima, S; Asada, Y; Kawabata, S. Clinical and bacteriologic evaluation of OPC-7251 in patients with acne: a double blind group comparison study vs cream base. *J M Acad Dermatol* 1991; 25: 674–681.

[146] Nishijima, S, Kurokawa, I; Nakaya, H. Susceptibility change to antibiotics of *Staphylococcus aureus* strains isolated from skin infections between July 1994 and November 2000. *J Infect Chemother* 2002; 8: 187–189.

[147] Nishijima, S; Nakagawa, M; Sugiyama, T; Akamatsu, H; Horio, T; Kawabata, S; Fujita, M. Sensitivity of *Staphylococcus aureus*, isolated

from skin infections in 1994 to 19 antimicrobial agents. *J Int Med Res* 1995; 23: 328–334.
[148] Morita, S; Otsubo, K; Matsubara, J; Ohtnai, T; Uchida, M. An efficient synthesis of a key intermediate towards (*S*)-(-)-nadifloxacin. *Tetrahedron: Asymmetry* 1995; 6: 245–254.
[149] Nishijima, S; Namura, S; Akamatsu, H; Kawai, S; Asada, Y; Kawabata, S; Fujita, M. In vitro activity of nadifloxacin against both methicillin-susceptible and -resistant clinical isolates of *Staphylococcus aureus* from patients with skin infections. *Drugs* 1995; 49(Suppl. 2): 230–232.
[150] De Souza, NJ; Gupte, SV; Deshpande, PK; Desai, VN; Bhawsar, SB; Yeole, RD; Shukla, MC; Strahilevitz, J; Hooper, DC; Bozdogan, B; Appelbaum, PC; Jacobs, MR; Shetty, N; Patel, MV; Jha, R; Khorakiwala, HF. A chiral benzoquinolizine-2-carboxylic acid arginine salt active against vancomycin-resistant *Staphylococcus aureus*. *J Med Chem* 2005; 48: 5232–5242.
[151] Patel, MV; De Souza, NJ; Gupte, SV; Jafri, MA; Bhagwat, SS; Chugh, Y; Khorakiwala, HF; Jacobs, MR; Appelbaum, PC. Antistaphylococcal activity of WCK 771, a tricyclic fluoroquinolone, in animal infection models. *Antimicrob Agents Chemother* 2004; 48: 4754–4761.
[152] Bhagwat, SS; Mundkur, LA; Gupte, SV; Patel, MV; Khorakiwala, HF. The anti-methicillin-resistant *Staphylococcus aureus* quinolone WCK 771 has potent activity against sequentially selected mutants, has a narrow mutant selection window against quinolone-resistant *Staphylococcus aureus*, and preferentially targets DNA gyrase. *Antimicrob Agents Chemother* 2006; 50: 3568–3579.
[153] Greka, P; Souli, M; Athanasiou, K; Mandaraka, A; Katsala, D; Bourousi, M; Giamarellou, H. 39th Interscience Conference on Antimicrobial Agents and Chemotherapy, 1999, Abstract No. E-1495.
[154] Goldstein, EJ; Conrads, G; Citron, DM; Merriam, V; Warren, Y; Tyrrell, K. In vitro activity of gemifloxacin compared to seven other oral antimicrobial agents against aerobic and anaerobic pathogens isolated from antral sinus puncture specimens from patients with sinusitis. *Diagn Microbiol Infect Dis* 2002; 42: 113–118.
[155] Fuchs, PC; Barry, AL; Brown, SD. In vitro activity of gemifloxacin against contemporary clinical bacterial isolates from eleven North American medical centers, and assessment of disk diffusion test interpretive criteria. *Diagn Microbiol Infect Dis* 2000; 38: 243–253.
[156] Blondeau, JM; Hansen, G; Metzler, KL; Borsos, S; Irvine, LB; Blanco, L. In vitro susceptibility of 4903 bacterial isolates to gemifloxacin—an advanced fluoroquinolone. *Int J Antimicrob Agents* 2003; 22: 147–154.

[157] Yun, HJ; Min, YH; Lim, JA; Kang, JW; Kim, SY; Kim, MJ; Jeong, JH; Choi, YJ; Kwon, HJ; Jung, YH; Shim, MJ; Choi, EC. In vitro and in vivo antibacterial activities of DW286, a new fluoronaphthyridone antibiotic. *Antimicrob Agents Chemother* 2002; 46: 3071–3074.

[158] Kim, MJ; Yun, HJ; Kang, JW; Kim, S; Kwak, JH; Choi, EC. In vitro development of resistance to a novel fluoroquinolone, DW286, in methicillin-resistant *Staphylococcus aureus* clinical isolates. *J Antimicrob Chemother* 2003; 51: 1011–1016.

[159] Kim, HJ; Seol, MJ; Park, HS; Choi, DR; Seong, SK; Shin, HK; Kwak, JH. Antimicrobial activity of DW-224a, a new fluoroquinolone, against *Streptococcus pneumoniae*. *J Antimicrob Chemother* 2006; 57: 1256–1258.

[160] Kwon, AR; Min, YH; Ryu, JM; Choi, DR; Shim, MJ; Choi, EC. In vitro and in vivo activities of DW-224a, a novel fluoroquinolone antibiotic agent. *J Antimicrob Chemother* 2006; 58: 684–688.

[161] Park, HS; Kim, HJ; Seol, MJ; Choi, DR; Choi, EC; Kwak, JH. In vitro and in vivo antibacterial activities of DW-224a, a new fluoronaphthyridone. *Antimicrob Agents Chemother* 2006; 50: 2261–2264.

[162] Kim EJ; Shin WH; Kim KS; et al. Safety pharmacology of DW-224a, a novel fluoroquinolone antibiotic agent. *Drug Chem Toxicol* 2004; 27: 295–307.

[163] Han, J; Kim, JC; Chung, MK; Kim, B; Choi, DR. Subacute toxicity and toxicokinetics of a new antibiotic, DW-224a, after single and 4-week repeated oral administration in dogs. *Biol Pharm Bull* 2003; 26: 832–839.

[164] Bush, K; Macielag, M; Weidner-Wells, M. Taking inventory: antibacterial agents currently at or beyond phase 1. *Curr Opin Microbiol* 2004; 7: 466–476.

[165] Christiansen, KJ; Bell, JM; Turnidge, JD; Jones, RN. Antimicrobial activities of garenoxacin (BMS 284756) against Asia-Pacific region clinical isolates from the SENTRY program, 1999 to 2001. *Antimicrob Agents Chemother* 2004; 48: 2049–2055.

[166] Fung-Tomc, JC; Minassian, B; Kolek, B; Huczko, E; Aleksunes, L; Stickle, T; Washo, T; Gradelski, E; Valera, L; Bonner, DP. Antibacterial spectrum of a novel des-fluoro(6) quinolone, BMS-284756. *Antimicrob Agents Chemother* 2000; 44: 3351–3356.

[167] Firsov, AA; Vostrov, SN; Lubenko, IY; Arzamastsev, AP; Portnoy, YA; Zinner, SH. ABT492 and levofloxacin: comparison of their pharmacodynamics and their abilities to prevent the selection of

resistant *Staphylococcus aureus* in an in vitro dynamic model. *J Antimicrob Chemother* 2004; 54: 178–186.

[168] Pfaller, MA; Beach, ML; Gordon, KA; Jones, RN. Comparative antimicrobial spectrum and activity of BMS284756 (T-3811; a desfluoroquinolone) tested against an international collection of staphylococci and enterococci, including in vitro test development and international comparisons. *J Chemother* 2002; 14: 13–18.

[169] Watanabe, S; Ito, T; Hiramatsu, K. Susceptibilities of healthcare- and community-associated methicillin-resistant staphylococci to the novel des-F(6)-quinolone DX-619. *J Antimicrob Chemother* 2007; 60: 1384–1387.

[170] Yamamoto, T; Dohmae, S; Saito, K; Otsuka, T; Takano, T; Chiba, M; Fujikawa, K; Tanaka, M. Molecular characteristics and in vitro susceptibility to antimicrobial agents, including the des-fluoro(6) quinolone DX-619, of Panton-Valentine leucocidin-positive methicillin-resistant *Staphylococcus aureus* isolates from the community and hospitals. *Antimicrob Agents Chemother* 2006; 50: 4077–4086.

[171] Bogdanovich, T; Esel, D; Kelly, LM; Bozdogan, B; Credito, K; Lin, G; Smith, K; Ednie, LM; Hoellman, DB; Appelbaum, PC. Antistaphylococcal activity of DX-619, a new des-F(6)-quinolone, compared to those of other agents. *Antimicrob Agents Chemother* 2005; 49: 3325–3333.

[172] Strahilevitz, J; Truong-Bolduc, QC; Hooper, DC. DX-619, a novel des-fluoro(6) quinolone manifesting low frequency of selection of resistant *Staphylococcus aureus* mutants: quinolone resistance beyond modification of type II topoisomerases. *Antimicrob Agents Chemother* 2005; 49: 5051–5057.

[173] Recommended International Nonproprietary Names: List 58, *WHO Drug Information* 2007; 21: 256.

[174] 48th Annual Interscience Conference on Antimicrobial Agents and Chemotherapy / 46th Annual Meeting of Infectious Diseases Society of America, 2008; Washington, D.C.

[175] Recommended International Nonproprietary Names: List 58, *WHO Drug Information* 2007; 21: 257.

[176] Yamakawa, T; Mitsuyama, J; Hayashi, K. In vitro and in vivo antibacterial activity of T-3912, a novel non-fluorinated topical quinolone. *J Antimicrob Chemother* 2002; 49: 455–465.

[177] Tran, TP; Ellsworth, EL; Stier, MA; Domagala, JM; Showalter, HDH; Gracheck, SJ; Shapiro, MA; Joannides, TE; Singh, R. Synthesis and structural–activity relationships of 3-hydroxyquinazoline-2,4-dione antibacterial agents. *Bioorg Med Chem Lett* 2004; 14: 4405–4409.

[178] Huband, MD; Cohen, MA; Zurack, M; Hanna, DL; Skerlos, LA; Sulavik, MC; Gibson, GW; Gage, JW; Ellsworth, E; Stier, MA; Gracheck, SJ. In vitro and in vivo activities of PD 0305970 and PD 0326448, new bacterial gyrase/topoisomerase inhibitors with potent antibacterial activities versus multidrug-resistant gram-positive and fastidious organism groups. *Antimicrob Agents Chemother* 2007; 51: 1191–1201.

[179] Ellsworth, EL; Tran, TP; Showalter, HD; Sanchez, JP; Watson, BM; Stier, MA; Domagala, JM; Gracheck, SJ; Joannides, ET; Shapiro, MA; Dunham, SA; Hanna, DL; Huband, MD; Gage, JW; Bronstein, JC; Liu, JY; Nguyen, DQ; Singh, R. 3-aminoquinazolinediones as a new class of antibacterial agents demonstrating excellent antibacterial activity against wild-type and multidrug resistant organisms. *J Med Chem* 2006; 49: 6435–6438.

[180] Nilius, AM; Hensey-Rudloff, DM; Almer, LS; Fung, A; Arminger, YL; Chu, D; Flamm, RK. Comparative in vitro antibacterial activity of A-170568.1, a novel 2-pyridone antibacterial agent. Abstract F79. 38th Interscience Conference on Antimicrobial Agents and Chemotherapy; 1998.

[181] Armiger, YL; Chu, DTW; Fung, AKL; Li, Q; Wang, W; Nilius, A; Alder, J; Ewing, P; Stone, G; Meulbroek, J; Bui, M; Shen, LL; Paige, L; Or, YS; Plattner, JJ. The discovery of A-165753 and A-170568, two potent broad spectrum antimicrobial agents. Abstract F86. 38th Interscience Conference on Antimicrobial Agents and Chemotherapy; 1998.

[182] Alder, J; Clement, J; Meulbroek, J; Shipkowitz, N; Mitten, M; Jarvis, K; Oleksijew, A; Hutch, T; Sr Paige, L; Flamm, R; Chu, D; Tanaka, K. Efficacies of ABT-719 and related 2-pyridones, members of a new class of antibacterial agents, against experimental bacterial infections. *Antimicrob Agents Chemother* 1995; 39: 971–975.

[183] Flamm, RK; Vojtko, C; Chu, DT; Li, Q; Beyer, J; Hensey, D; Ramer, N; Clement, JJ; Tanaka, SK. In vitro evaluation of ABT-719, a novel DNA gyrase inhibitor. *Antimicrob Agents Chemother* 1995; 39: 964–970.

[184] Chin, NX; Chu, D; Neu, HC. In vitro antibacterial activity of A-86719.1, a new class of DNA gyrase inhibitor. Abstract F49. 34th Interscience Conference on Antimicrobial Agents and Chemotherapy; 1994.

[185] Robertson, GT; Bonventre, EJ; Doyle, TB; Du, Q; Duncan, L; Morris, TW; Roche, ED; Yan, D; Lynch, AS. In vitro evaluation of CBR-2092, a novel rifamycin-quinolone hybrid antibiotic: studies of the mode of

action in *Staphylococcus aureus*. *Antimicrob Agents Chemother* 2008; 52: 2313–2323.

[186] Li, Q; Mitscher, LA; Shen, LL. The 2-pyridone antibacterial agents: bacterial topoisomerase inhibitors. *Med Res Rev* 2000; 20: 231–293.

[187] Artsimovitch, I; Vassylyeva, MN; Svetlov, D; Svetlov, V; Perederina, A; Igarashi, N; Matsugaki, N; Wakatsuki, S; Tahirov, TH; Vassylyev, DG. Allosteric modulation of the RNA polymerase catalytic reaction is an essential component of transcription control by rifamycins. *Cell* 2005; 122: 351–363.

[188] Campbell, EA; Korzheva, N; Mustaev, A; Murakami, K; Nair, S; Goldfarb, A; Darst, SA. Structural mechanism for rifampicin inhibition of bacterial RNA polymerase. *Cell Microbiol* 2001; 104: 901–912.

[189] Robertson, GT; Bonventre, EJ; Doyle, TB; Du, Q; Duncan, L; Morris, TW; Roche, ED; Yan, D; Lynch, AS. In vitro evaluation of CBR-2092, a novel rifamycin-quinolone hybrid antibiotic: microbiology profiling studies with staphylococci and streptococci. *Antimicrob Agents Chemother* 2008; 52: 2324–34.

[190] Emami, S; Shafiee, A; Foroumadi, A. Structural features of new quinolones and relationship to antibacterial activity against Gram-positive bacteria. *Mini-Rev Med Chem* 2006; 6: 375–386.

[191] Nseir, S; Ader, F; Marquette, CH; Durocher, A. Impact of fluoroquinolone use on multidrug-resistant bacteria emergence. *Pathol Biol (Paris)* 2005; 53: 470–475.

[192] John, JF; Harvin, AM. History and evolution of antibiotic resistance in coagulase-negative staphylococci: Susceptibility profiles of new anti-staphylococcal agents. *Ther Clin Risk Manag* 2007; 3: 1143–1152.

[193] Chide, OE; Orisakwe, OE. Structural development, haematological immunological and pharmacological effects of quinolones. *Recent Pat Antiinfect Drug Discov* 2007; 2: 157–168.

[194] De Souza, MV. New fluoroquinolones: a class of potent antibiotics. *Mini-Rev Med Chem* 2005; 5: 1009–1017.

INDEX

A

absorption, 22
acid, xiii, 1, 5, 13, 17, 18, 20, 27, 32, 44, 48, 54, 55, 63, 66, 70, 76, 77, 80, 100
acne, 99
administration, 32, 48, 55, 56, 69, 101
aerobic, 100
agent, xiii, 1, 14, 49, 54, 56, 76, 101, 103
agents, xiii, xiv, 5, 15, 16, 22, 23, 62, 71, 77, 85, 87, 100, 102, 103, 104
allele, 97
alternative, 18
alveolar macrophage, 48, 99
amine, 22, 23, 55, 69, 76, 77, 80
amino, 13, 19, 21, 22, 27, 32, 44, 48, 54, 76, 78, 80, 93, 94, 95, 97
amino acid, 13
anaerobic, 100
analog, 18, 20, 33, 63, 78
animal models, 61
animals, 48
antibacterial agents, xi, xii, xiii, 71, 78, 79, 94, 99, 101, 102, 103, 104
antibiotic, xiii, 11, 15, 54, 101, 103, 104
antibiotic resistance, xiii, 104
antibiotics, 1, 11, 70, 99, 104
antimycobacterial activity, 95
antineoplastic, 94
antineoplastic agents, 94
arginine, 55, 56, 100
assessment, 100
ATP, 8
attachment, 8, 77
avoidance, 16

B

bacteria, xiii, 7, 8, 13, 14, 15, 16, 17, 18, 21, 22, 25, 28, 30, 33, 44, 48, 49, 54, 55, 66, 81, 98, 104
bacterial, xi, 1, 7, 8, 12, 15, 22, 78, 83, 88, 95, 100, 103, 104
bacterial infection, xi, 83, 103
bacteriostatic, 8, 25, 97
bacterium, 12
barrier, 23, 48
binding, 17, 18, 19, 32, 48, 88, 90
bioavailability, 22
biofilms, 82

biological activity, xi, 17, 18, 85
blood, 5
body weight, 69
bonding, 8
breakdown, 8
broad spectrum, xiii, 2, 33, 44, 54, 103
bromine, 25
bronchitis, 55

C

carbon, 62, 69, 77
carbonyl groups, 17
carboxyl, 8, 19
carboxylic, 17, 18, 44, 48, 54, 55, 63, 66, 70, 76, 80, 87, 94, 99, 100
carboxylic acids, 87, 94, 99
catheters, 82
cell, 1, 8, 14, 20, 22, 48
cell growth, 8
cell membranes, 48
cellulitis, 61
central nervous system, 5, 48
chemotherapy, 98, 99, 100, 102, 103
chiral, 93, 100
chlorine, 25
chloroquine, 1
chromosome, 88, 98
ciprofloxacin, xiii, 5, 8, 9, 11, 12, 13, 14, 15, 17, 20, 21, 22, 30, 32, 33, 44, 47, 48, 49, 54, 56, 62, 63, 64, 66, 67, 68, 69, 70, 71, 80, 81, 82, 83, 85, 90, 91, 92, 93, 95
classes, 1, 15
classical, 12
clinical trial, 55, 82
colonization, 11
community, 33, 42, 43, 55, 70, 71, 76, 102
complement, 7
complementary DNA, 8
compounds, xi, xii, xiii, 1, 5, 17, 20, 21, 22, 27, 30, 49, 82, 85, 95, 96, 98

confidence interval, 37
configuration, 62
control, 104
covalent, 8
cytoplasmic membrane, 23
cytotoxicity, 25

D

defects, 5
deficit, 18
definition, 5
derivatives, 14, 18, 23, 87, 96, 97, 98
diarrhea, 1
diffusion, 23, 100
dimer, 16
distribution, 15, 98
DNA, xi, 7, 8, 9, 12, 13, 14, 17, 18, 22, 61, 62, 77, 83, 88, 89, 90, 91, 100, 103
DNA sequencing, 90
dogs, 69, 101
drug action, 88
drug efflux, 14
drug resistance, xi, xiii, 62
drug use, 70
drug-resistant, xi, 24, 25, 85, 93
drugs, xi, xiii, 1, 5, 8, 12, 23, 33, 49, 63, 71
dyes, 15

E

efflux mechanisms, 92
enantiomers, 94
energy, 8
enterococci, 102
enzyme interaction, 8
enzymes, 7, 13, 77
epithelial lining fluid, 48, 99
equilibrium, 95
estimating, 92
evolution, 104

Index

excretion, 32
exporter, 62
exposure, 15

F

family, 1
flexibility, 22, 23
fluid, 48, 99
fluorinated, 9, 13, 18, 20, 48, 69, 77, 89, 92, 95, 102
fluorine, xiii, 1, 5, 20, 21, 69, 76
fluoroquinolones, xi, xii, xiii, 3, 5, 15, 20, 25, 27, 28, 33, 62, 69, 85, 87, 88, 89, 90, 91, 92, 93, 94, 96, 99, 104
furan, 96

G

gastrointestinal tract, 5
gene, 14, 90, 91
gene promoter, 14
generation, 3, 4, 5
genes, 7, 71
gentamicin, 77
gram-negative, xiii, 1, 5, 8, 20, 22, 30, 66, 79, 80, 85
gram-positive, xiii, 1, 5, 8, 13, 15, 16, 17, 18, 20, 21, 22, 23, 25, 28, 30, 33, 44, 48, 49, 55, 62, 66, 70, 79, 80, 81, 88, 89, 104
groups, 5, 17, 19, 22, 23, 25, 33, 63, 103
growth, 7, 8, 97

H

halogen, 19, 25, 85, 97
halogens, 25, 28
healthcare, xiii, 70, 102
hospitalized, 11
hospitals, xiii, 11, 70, 71, 102

humans, 32
hybrid, 103, 104
hybrids, 23, 24, 82, 97
hydrate, 27
hydro, 15
hydrochloric acid, 63
hydrogen, 8, 18, 20
hydrophilic, 15
hydrophobic, 8, 15, 17, 80
hydrophobic interactions, 8
hydrophobicity, 15, 44, 48, 96
hydroxyl, 18

I

immunocompromised, 49
immunological, 104
in vitro, 11, 15, 19, 25, 28, 44, 45, 47, 49, 54, 63, 66, 71, 76, 77, 81, 83, 89, 90, 93, 95, 97, 99, 102, 103
in vivo, 49, 63, 66, 76, 82, 89, 95, 98, 101, 102, 103
incidence, 55
indication, 1
infections, xi, xiii, xiv, 1, 5, 33, 48, 49, 54, 61, 62, 64, 69, 70, 78, 80, 82, 83, 88, 98, 99, 100, 103
inhibition, 7, 9, 12, 20, 88, 89, 95, 104
inhibitors, 15, 78, 80, 81, 93, 103, 104
inhibitory, 8, 77, 82
injection, 56
interaction, 9, 83
interactions, 8, 91
interference, 19
intravenous, 55
intravenously, 48

L

likelihood, 62
liniment, 55

linkage, 77
lipophilic, 55
lipopolysaccharide, 23
liver, 61
locus, 91, 98
low-level, 14, 15, 92
lung, 48, 61, 99

M

macrophage, 48, 99
market, 18
metabolism, xi
methicillin-resistant, xi, 33, 63, 71, 90, 92, 100, 101, 102
methyl group, 18, 19, 21, 22, 44, 49, 63, 81
mice, 32, 37, 49, 61, 64, 69, 99
microbial, 88
microorganisms, 20, 54
models, 32, 37, 49, 61, 100
moderate activity, 20
modulation, 104
moieties, 5, 23, 97
molecular mass, 23, 96
molecular structure, 91
molecules, 5, 8, 15, 77
monkeys, 32, 98
monomeric, 7
mouse, 61, 66, 80
mouse model, 66
mRNA, 7, 14
murine model, 49
murine models, 49
mutant, 9, 25, 49, 66, 71, 89, 93, 97, 100
mutants, 8, 12, 14, 15, 25, 49, 55, 61, 69, 71, 76, 77, 89, 92, 93, 100, 102
mutation, 11, 13, 16, 71, 77, 90
mutation rate, 11
mutations, 8, 9, 12, 13, 14, 61, 62, 71, 76, 90, 91
mycobacteria, 97

mycobacterium, 97

N

nitrogen, 17, 25, 55, 62
norfloxacin, 2, 5, 9, 15, 18, 20, 22, 92, 93
nucleus, 55, 85

O

ofloxacin, xiii, 9, 15, 18, 20, 22, 32, 35, 36, 44, 47, 48, 49, 50, 55, 62, 77, 78, 91, 94
omeprazole, 15
optimization, 20, 22, 23
oral, 22, 25, 32, 44, 48, 54, 56, 61, 69, 98, 100, 101
organism, 103

P

partition, 56
pathogenic, 54
pathogens, 20, 22, 25, 33, 44, 49, 69, 82, 89, 100
patients, xiii, 11, 33, 63, 71, 99, 100
permeability, 48, 56
pharmacodynamics, 101
pharmacokinetics, xiii, 22, 25, 98
pharmacological, 62, 104
pharmacology, 87, 101
phospholipids, 48
physicochemical, 15, 22, 56
physicochemical properties, 15, 56
physiological, 88
pigs, 48
play, 25
pneumonia, 55
point mutation, 90
polymerase, 7, 82, 90, 104
polymerase chain reaction, 90

poor, 66
population, 90
preclinical, xi, xiv, 85
preference, 9, 12, 61, 89, 92
prodrugs, 56
program, 82, 101
promoter, 14
prosthetic device, 82
protection, 62, 80
protein, 14, 15, 32, 90
protein binding, 32
protein synthesis, 14
proteins, 14, 23, 62
pseudomonas, 1, 5, 89, 92, 96
pseudomonas aeruginosa, 5, 89, 92, 96
pumps, 14, 15, 66, 76
PVL, 71

R

race, 55, 62
radiolabeled, 32
range, 11, 33, 61
rat, 80
rats, 32, 69, 98, 99
reality, 93
refractory, 16
relationship, 23, 91, 98, 104
relationships, xi, xiv, 17, 87, 91, 93, 94, 102
replication, xi, 7, 88
residues, 96
resistance, xi, xiii, 1, 5, 6, 11, 12, 13, 14, 15, 16, 49, 61, 62, 66, 71, 76, 77, 87, 88, 90, 91, 92, 93, 95, 97, 98, 99, 101, 102, 104
respiratory, 1, 5, 44, 54, 69, 88, 98
rings, 22
risk, 25, 28
RNA, 7, 82, 104

S

safety, 6, 69, 76, 94
salt, 55, 56, 63, 81, 100
salts, 56, 81
SAR, 17, 18, 85, 95
scaffold, 44, 83
second generation, 1
selectivity, 92, 95
separation, 18
septicemia, 32, 37
serine, 13
serum, 1
shares, 49, 70
side effects, 94
similarity, 49, 70
sinus, 100
sinusitis, 63, 100
sites, 48
skeleton, 1
skin, 69, 78, 99, 100
solubility, 56
spatial, 17, 20, 70, 85
species, 1, 5, 8, 12, 54
spectrum, xiii, 1, 5, 20, 22, 25, 30, 33, 44, 54, 55, 89, 95, 101, 102, 103
spleen, 61
stability, 14
stabilization, 89
staphylococcal, xi, xiv, 4, 22, 23, 54, 56, 61, 104
staphylococci, xiii, 13, 18, 23, 25, 33, 40, 45, 47, 52, 54, 55, 56, 59, 62, 63, 66, 68, 71, 74, 75, 76, 78, 79, 82, 83, 85, 102, 104
staphylococcus aureus, xi, xiii, 79, 87, 88, 89, 90, 91, 92, 93, 96, 97, 98, 99, 100, 101, 102, 104
steric, 17, 18
strain, 9, 48, 61

strains, xi, xiv, 9, 11, 12, 15, 16, 20, 24, 25, 30, 31, 32, 33, 49, 54, 55, 56, 61, 62, 63, 66, 69, 70, 71, 77, 85, 91, 93, 99
streptococci, 33, 62, 83, 104
stress, 7
structural characteristics, 15
structural gene, 14
structural modifications, 77
substitution, 5, 14, 19, 20, 22, 76
substrates, 15, 76
survival, 14, 49, 97
susceptibility, 8, 11, 13, 15, 33, 90, 100, 102
synthesis, 1, 7, 8, 14, 97, 100

T

targets, 7, 8, 9, 12, 13, 17, 18, 61, 89, 100
teicoplanin, xiii, 49
ternary complex, 8
tissue, 69, 78, 87, 98
TMA, 95
tolerance, 23
topological, 7
toxicity, 1, 101
toxicological, 48
transcription, 7, 14, 104
transmission, 11
transport, 15

U

urinary, 1, 69
urinary tract, 1, 69
urinary tract infection, 1

V

values, 8, 23, 63, 71, 79, 81
vancomycin, xiii, 49, 61, 71, 76, 80, 82, 99, 100
variation, 17, 18, 23

W

water, 56, 81
water-soluble, 56
WHO, 102